Regional Fisheries Management Organizations

Leandra R. Gonçalves

Regional Fisheries Management Organizations

The Interplay between Governance and Science

Leandra R. Gonçalves
Praça do Oceanográfico
University of Sao Paulo
São Paulo, Brazil

ISBN 978-3-030-70361-5   ISBN 978-3-030-70362-2   (eBook)
https://doi.org/10.1007/978-3-030-70362-2

This Springer imprint is published by the registered company Springer Nature Switzerland AG
The registered company address is: Gewerbestrasse 11, 6330 Cham, Switzerland

"Gonçalves provides an accessible and comprehensive analysis of RMFOs. She offers valuable insights into the role of science and politics in shaping sustainable fisheries policies for the open oceans."

—Peter M. Haas, *Professor Department of Political Science, University of Massachusetts Amherst*

"It is highly emblematic that the book *Regional Fisheries Management Organizations: The interplay between governance and science*, authored by Leandra Gonçalves, is coming to life precisely in the first year of the Decade of Ocean Science for Sustainable Development (2021–2030), proclaimed by the United Nations. This is exactly what this book is all about: how and, more importantly, WHEN knowledge and science can influence for the better the political decisions required to ensure the sustainable development of our oceans. Focused on the ability of Regional Fisheries Management Organizations to conserve and manage highly migratory fish stocks, the book addresses the main threat to the future of marine biodiversity: unsustainable fisheries management, by posing a question that cuts to the heart of the problem: when does policy listen to science? By using the epistemic communities' theory, the author arrives to the conclusion that science may indeed speak louder to policy, but the strategy, allies, and the ways in which it occurs differ case-by-case. As envisaged by the UN proclamation, this book is an important and sincere brick, hopefully to be accompanied by many others to come during this promising decade, which will help to build a common framework to ensure that ocean science can support countries and the international community in creating improved conditions for the sustainable development of our cherished Oceans."

—Fabio H. Hazin, *Professor at Federal Rural University of Pernambuco, Brazil*

To Dante, my son, and his generation, which I hope will believe that knowledge may change politics for a better world.

Foreword

When I attended my first regional fisheries management organization (RFMO) meeting as an observer in 2001, I was a firm believer in instrumentally rational interpretations of state behavior. Indeed, much of my work on international fisheries governance is clearly based on the liberal-institutionalist school. However, through long experience with RFMOs I also gained an appreciation for the constructivist approach and the role of epistemic communities in international environmental governance. While I still believe that the structure of power and interests determines the outcomes of negotiations, I have also seen firsthand how states' understanding of both is shaped by social interactions.

In this book, Leandra Gonçalves advances our understanding of the construction of state preferences by examining how science succeeds or fails to influence decisions by three regional fisheries management organizations: The Commission for the Conservation of Antarctic Marine Living Resources (CCAMLR), the International Commission for the Conservation of Atlantic Tunas (ICCAT), and the Commission for the Conservation of Southern Bluefin Tuna (CCSBT). These RFMOs provide interesting cases because their formal scientific advisory mechanisms are similar, but they differ in the cohesiveness and legitimacy of associated epistemic communities on one hand and the "usefulness" of the scientific advice that they can provide on the other. Furthermore, she demonstrates that the influence of "useful knowledge" is context dependent and that relationships between epistemic communities and other factors used to explain state behavior are quite complex.

For instance, CCAMLR is well known as one of the most conservation-oriented RFMOs, and member states have evinced strong willingness to act on advice from its scientific committee. However, high levels of illegal, unregulated, and unreported fishing have made it impossible for the scientific community to collect the data needed to accurately assess fish stocks. This makes it difficult for the scientific committee to produce useable knowledge, which in turn prevents CCAMLR from adopting management measures based on scientific advice.

In contrast, ICCAT is considered to have one of the worst conservation records in international fisheries management and their scientific committee is known for being highly politicized. Furthermore, ICCAT frequently chooses to set total allowable catches above scientifically recommended levels and otherwise ignore scientific advice. However, in 2009, an epistemic community coalesced around eastern bluefin tuna produced usable knowledge that clearly indicated an impending crisis for the stock and helped to propel the commission toward management that conformed more readily to scientific advice. As Gonçalves notes, the crisis faced by ICCAT was not just ecological but also political and economic, as both public pressure and a potential listing of eastern bluefin tuna by the Convention on the International Trade in Endangered Species (CITES) provided additional incentives to adopt stricter conservation measures for the stock. It is not possible to determine which of these factors was most important, but Gonçalves asserts that all three were necessary to effect the change in ICCAT's management of eastern bluefin tuna.

For her third example, Gonçalves shows how a major crisis in data can foster changes in formal rules regarding scientific advisory bodies and, eventually, shifts in informal norms regarding the legitimacy of scientific advice. In the mid-2000s, members of the CCSBT discovered that one of the largest fishing states in the regime had been substantially underreporting its catches for decades. This threw the scientific community into disarray and caused considerable conflict over how to account for the missing data in stock assessments. This lack of consensus continued until the commission adopted new rules governing the production of scientific advice. This, in turn, helped reduce conflict and eventually helped re-establish trust in scientific advice.

Demonstrating the causal relationships between epistemic communities and the outcomes of international negotiations is a difficult, time intensive process that often requires expert interviews, observation, and analysis of primary documents. This makes it an expensive undertaking, but as Gonçalves shows, it is important if we are to move beyond the false dichotomies created by our ideological blinders to understand not whether but when epistemic communities influence international negotiations. Oran Young pointed out the need for such a "third path" many years ago, but progress has been slow. Gonçalves takes us one more step down that road.

Environmental Studies Program   D. G. Webster
Dartmouth College
Hanover, NH, USA

Abstract

One of the biggest challenges in contemporary global environmental governance is the future of marine biodiversity. Over the years, increased fishing efforts in previously remote areas drove many fish stocks to scarcity.

The Regional Fisheries Management Organizations (RFMOs) emerged to solve the international fishery crisis, on the assumption that they would provide a forum where Member States may agree and discuss binding rules for the conservation and management of fish stocks within its geographical area of responsibility.

Although some agreements existed for more than 60 years, many authors agree that they have not been fully effective in promoting the maintenance and conservation of fish stocks. There are many reasons that might explain the lack of effectiveness, one of them being that science is not very often listened to in the political decision-making process.

In this research, the influence of knowledge and science in shaping policy decisions will be observed and analyzed. Therefore, the theory of epistemic communities—which forms part of the constructivist turn in international relations—was used to answer the main question posed here: when does power listen to science? When it does, does it bring more effectiveness in terms input from epistemic communities? Does it induce states to change their behavior, and do these lead to policies, which can credibly improve biomass?

Using process tracing, through elite interviews, and with a systematic compilation and study of meeting reports from three RFMOs: the Commission for the Conservation of Antarctic Marine Living Resources (CCAMLR); the International Commission for the Conservation of Atlantic Tunas (ICCAT); and the Commission for the Conservation of Southern Bluefin Tuna (CCSBT), the final conclusion is that the science may speak louder to policy, but the strategy, allies, and the ways in which it occurs differ case-by-case.

Keywords Fishery resources; International organizations; Epistemic Communities; Oceans

Acknowledgments

This book is a result of my doctoral thesis concluded in 2017 in International Relation Institute at the University of São Paulo. I am glad to see it published as this would not have been possible without the advice and support of many people.

Those who deserve warm thanks for their help during the preparation of this book are indeed many. It is not an easy task to find your place in the academic realm when you want to combine the natural world with social science. I could not cope had it not been for friends, family, and my advisor's companionship.

My advisor **Dr. José Eli da Veiga**, who relied on my work, patiently reviewed all my materials, supported all my initiatives, and taught me that there would be nothing more important than a good research question and its sharp hypothesis. And more than that to conduct a research that gives us nothing but pleasure. **Dr. Peter Haas**, who welcomed me to the University of Massachusetts for almost a year, was always very gentle and available to help me to show the right pathway for my research, who kept me focused as much as I could and who assisted me anytime I needed it, even on my existential doubts, which I learned from him are entirely appropriate—and we will never lose them.

For my mentors at IR studies **Professor Janina Onuki, Professor Maria Hermínia Tavares de Almeida, and Professor Eduardo Viola**, who were responsible for introducing me to the IR realm—the new theoretical world I learned and I am still navigating in shallow waters.

Next in line is a person who has had the greatest influence on my thinking about Regional Fisheries Management Organizations and how, even with so many difficulties, they can help to improve the international fisheries management: **Dr. Fabio Hazin**.

To my colleagues at the International Studies Association and at the Earth System Governance Project, where I have presented preliminary versions of this research, who have given me firm encouragement, candid criticism, and a keen eye for the larger picture: **Dr. DG Webster, Dr. Elizabeth De Sombre, Dr. Samuel Barking, Dr. Oran Young, and Dr. Ana Flavia Barros-Platiau** to name just a few.

I would like to thank **Professor Ilana Wainer**, a very special friend, who advised me from the beginning to look for International Relations PHd, even being a

Professor from the Oceanographic Institute. She pointed me a path that I have not foreseen before, and that now, I recognize it could not be different.

I am grateful to CAPES and Fulbright for providing the scholarship for most part of this work, and even more for creating the opportunity for going abroad, learning, and sharing my thoughts with a variety of other researchers. I am thankful for the "Santander mobilidade" grant. These certainly contributed much to the development of this study.

In exploring these new horizons and challenges, I have had the pleasure of being advised and supported by many outstanding researchers and colleagues in a variety of research and policy contexts. Most of these colleagues deserve special acknowledgement from my heart, and they know who they are. The list of colleagues and friends would be endless here.

I could not end the acknowledgements section without recognizing the role of my family and special friends who were with me throughout the Ph.D. period. My closest family: Marilena, Eduardo, and Samuel Gonçalves. From them, throughout my entire life, I had the strongest support, love, and the values that provided the courage to rely on my skills and to follow all of my dreams, independently, no matter how challenging they were.

My heartfelt thanks, however, go to my loving partner, Pedro Henrique Campello Torres—my most valued critic, strongest supporter, and closest friend throughout all these years. Our family may not experience the upcoming benefits of an effective agreement that will promote sustainable fisheries and a healthy ocean; however, hopefully this will be the future of ocean we will leave for future generations, and our son, Dante Gonçalves Torres.

Introduction

The importance of discussing the governance challenges in the Anthropocene goes through different dimensions within various ecosystems and biomes. The future of marine ecosystems is one of the priority areas to discuss sustainability, as not only they have been threatened at an unprecedented scale but they also provide humanity with different ecosystem services that guarantee our life on Earth, such as food provisioning, temperature control, water purification, coastal protection, prevention of erosion, water purification, and carbon storage – to name a few.

There are currently many agreements and institutional arrangements oriented to control the increasing pressure on the marine environment. However, we still observe a high rate of degradation of the marine environment, both within and beyond areas of national jurisdiction, which raises questions about the effectiveness of the regulatory framework in place. The main questions are: How can it be improved? Why are some of the agreements already in place failing to restore biodiversity and maintain its ecological process? How can research contribute to filling the deep vacuum concerning the role of the global governance of the oceans in the configuration and dynamics of contemporary international relations?

To address these questions, this research began with looking at the main regulatory instrument of this cooperation: the United Nations Convention on the Law of the Sea (UNCLOS), signed in 1982. This crucial document, known as the "Constitution of the Oceans," is the broadest and most complex text negotiated, to date, on this subject.

Nonetheless, UNCLOS has become an essential regulatory framework that generates diverse agreements towards other ocean's problems. Examples of it are regional fisheries agreements, regional seas agreements, and many others.

Thus, for the purpose of this book, keeping on the horizon the determination to understand the complex global governance of the oceans, it seemed more appropriate for developing research with specific case studies. Many issues could be considered for research, such as pollution, biodiversity, and acidification. However, international fisheries were the most interesting topic due to my familiarity with the subject, its intriguing complexity, and many other reasons explained below.

UNCLOS was negotiated under the *"mare liberum"* doctrine (Grotius 1604) setting the ocean free to every country. There was no sovereignty over the oceans. The freedom of the seas would be paramount in communication and trade between people and nations. No country should control the oceans, given its immensity and lack of limits. It should be free for fishing and navigation. And still, with that in mind, natural resources have been impacted and showing clear signs of declining.

With time, technology evolved and made fishing vessels faster and more powerful to fish more in less time. Over the years, increased fishing efforts in previously remote areas drove many fish stocks to scarcity.

For several decades, world fisheries have become a market-driven, dynamically developing sector of the food industry with large investments in modern fishing fleets and processing factories to meet the international demand for fish and fishery products. Since the 1960s, it has become apparent that fishery resources could no longer sustain such rapid and often uncontrolled increases in exploitation and development. Therefore, new approaches to fisheries management embracing conservation and environmental considerations were urgently needed.

The situation was aggravated by the realization that illegal, unregulated, and unreported fisheries on the high seas, in some cases involving straddling and highly migratory fish species, which occur within and outside Exclusive Economic Zones (EEZs), were impeding sound resource management (de Bruyn et al. 2013; Lindley and Techera 2017).

Most of these fisheries are considered international, meaning highly migratory species that move across international waters and different EEZs. Individual countries cannot effectively regulate and manage such fisheries. These fisheries are often referred to as common-pool resources (CPRs).[1] And thus, to promote effective management of international fisheries coordination among the State is needed. The Regional Fisheries Management Organizations (RFMOs) emerged and included the major world powers to play a central role in regulating the international fisheries.

Formal cooperation between states through RFMOs dates back to the early twentieth century but increased more rapidly from the 1960s. There are 38 regional fishery bodies worldwide: 20 advisory bodies and 18 RFMOs (Table 1). The FAO (2001, para 6, c[2]) defines RFMOs as "intergovernmental fisheries organizations or arrangements, as appropriate, that have the competence to establish fisheries conservation and management measures." Some of these, such as the International Whaling Commission (IWC) and the North Atlantic Salmon Conservation Organization (NASCO), have particular mandates or deal with single species. Others have broader mandates.

---

[1] Common-pool resources (CPRs) are natural resources where one person's use subtracts from another's use and where it is often necessary, but difficult and costly, to exclude other users outside the group from using the resource. The majority of the CPR research to date has been in the areas of fisheries, forests, grazing systems, wildlife, water resources, irrigation systems, agriculture, land tenure and use, social organization, theory (social dilemmas, game theory, experimental economics, etc.), and global commons (climate change, air pollution, transboundary disputes, etc.).

[2] http://www.fao.org/docrep/003/y1224e/y1224e00.htm#INTRODUCTION.

Table 1 Regional Fisheries Management Organizations (Lodge 2007)

		Year established
CCAMLR	Convention on the Conservation of Antarctic Marine Living Resources	1982
CCBSP	Convention on the Conservation and Management of the Pollock Resources in the Central Bering Sea	1996
CCSBT	Convention for the Conservation of Southern Bluefin Tuna	1994
GFCM	General Fisheries Council for the Mediterranean	1952
IATTC	Inter-American Tropical Tuna Commission	1950
IBSFC	International Baltic Sea Fisheries Commission	1973
ICCAT	International Convention for the Conservation of Atlantic Tunas	1969
IOTC	Indian Ocean Tuna Commission	1996
IPHC	International Pacific Halibut Commission	1923
IWC	International Whaling Commission	1946
NAFO	Northwest Atlantic Fisheries Organization	1979
NASCO	North Atlantic Salmon Conservation Organization	1983
NEAFC	North-East Atlantic Fisheries Commission	1982
NPAFC	North Pacific Anadromous Fish Commission	1993
PSC	Pacific Salmon Commission	1985
SEAFO	South East Atlantic Fisheries Organization	2003
SIOFA	South Indian Ocean Fisheries Agreement	2006
WCPFC	Western and Central Pacific Fisheries Commission	2004

Since 2003, new RFMOs have been established for the Western and Central Pacific Ocean (WCPFC), South-East Atlantic (SEAFO), and South Indian Ocean (SIOFA). A process is also underway to establish an RFMO for the Southern Pacific Ocean. Thus, while some critical gaps remain, both in terms of species and area coverage, most of the world's marine fish resources are now under management by one or more RFMOs (Lodge 2007).

Coastal states have raised interest in regulating and organizing marine living resources' uses through the establishment of quotas to avoid the risk of a reduction in fish stocks in the future. Although some agreements have existed for more than 60 years, many authors agree that they have not been fully effective in promoting fish stocks' maintenance and conservation (Cullis-Suzuki and Pauly 2010; Gjerde et al. 2013; Barkin and Desombre 2013).

Global capture fisheries production in 2018 reached 96.4 million tonnes, increasing 5.4% from the average of the previous 3 years (FAO 2020). The percentage of stocks fished at biologically unsustainable levels increased from 10% in 1974 to 34.2% in 2017[3] (FAO 2020).

---

[3] By FAO (2020) definition, stocks fished at biologically unsustainable levels have an abundance lower than the level that can produce the MSY and are being overfished. These stocks require strict management plans to rebuild stock abundance to full and biologically sustainable productivity. The stocks fished within biologically sustainable levels have abundance at or above the level associated with MSY. Stocks fished at the MSY level produce catches that are at or very close to their maximum sustainable production. Therefore, they have no room for further expansion in catches, and

There are many issue areas that could be chosen as an empirical subject for a study about which factors are conferring more or less effectiveness on the international agreements, as have previously been made by a diverse array of researchers. These subjects include the Ozone Treaty, the Baltic and North Seas Agreements, Acid Rain in Europe, the Convention on Long-range Transboundary Air Pollution, the International Whaling Commission, and many others addressed by these books and articles (Haas et al. 1995; Young 1999; Miles et al. 2002; Mitchell 2003).

However, the fishery agreements were chosen as they deal with a highly technical and transboundary issue that needs to be addressed by multilateral organizations facing a high degree of uncertainty. In addition, as they cover different regions, it will allow for a comparison between them, even considering that they were built in different contexts and were instituted by other countries.

Thus, one of the requirements is to understand why some of these fishery agreements are less effective than others and enhance their effectiveness. Without knowing these reasons, one can hardly improve the complex global governance of the oceans. This is the core motivation for this research, which will seek to highlight precisely one of the factors that might help enhance the RFMO's effectiveness.

Other researchers have also dedicated their thoughts to contribute to this actual debate about RFMO effectiveness. Gjerde et al. (2013) highlight some potential reasons why RFMOs have struggled thus far in carrying out their mandates. According to them, first, most RFMOs comprise states with interests in enhancing or maintaining their domestic fishing opportunities,[4] leading to the pursuit of short-term gains over long-term sustainable fishing.

Second, RFMOs and their Member States suffer few consequences for poor performance or overfishing, other than possibly lost fishing opportunities in the remote future. Distant Water Fishing Nations (DWFN) have, in the past, been able to swiftly shift to more fertile grounds. Outside of the compliance mechanisms of RFMOs such as blacklists, few, if any, penalties exist at the international level for failing to follow UN fisheries resolutions and best practice standards such as the Food and Agricultural Organization (FAO) Code of Conduct for Responsible Fisheries.[5] However, the FAO's Port State Measures Agreement should have sufficient states vote to ratify it, which could help to fill this gap.[6]

---

effective management must be in place to sustain their MSY. The stocks with a biomass considerably above the MSY level (underfished stocks) have been exposed to relatively low fishing pressure and may have some potential to increase their production.

[4] A few select fisheries or marine living resource organizations contain non-fishing states, e.g., CCAMLR, IWC, and Inter-American Tropical Tuna Commission (IATTC). This factor will be examined in the CCAMLR chapter.

[5] http://www.fao.org/docrep/005/v9878e/v9878e00.htm.

[6] Food and Agriculture Organization of the United Nations, Agreement on Port State Measures to Prevent, Deter and Eliminate Illegal, Unreported and Unregulated Fishing (2009), http://www.fao.org/fishery/topic/166283/en. This measure needs to be ratified by 25 states to come into force but has only been ratified by nine states thus far (last update accessed on August 2015—http://www.fao.org/fileadmin/user_upload/legal/docs/037s-e.pdf).

Third, the paradigm of regional institutions may be ill suited to a globalized world. Many RFMOs may have originated from a small number of states with a shared dependence on, vested interest in, a shared resource (likely a fish stock adjacent to or straddling their EEZ and a small number of DWFNs). Yet in the current world of global fisheries, a vessel fishing the high seas may fly a flag from one state, hire a captain from another, a crew from several more, with ownership held by a mostly stateless multinational corporation that may belong to holding companies in one or more jurisdictions. Further, vessels that fish in the high seas can fish a resource to local commercial extinction and then move on, not feeling the local depletion's effect.

And finally, most RFMOs maintain the position that the Member States are allowed to fish unless they reach an agreement (generally achieved via consensus) not to fish or restrict fishing. This creates a perverse incentive not to reach fisheries agreements since any agreement would limit a state's total allowable catch (TAC) and hence their "freedom to fish." Consequently, fishing limits are often only adopted following the stock collapse or after severe environmental impacts have ensued (e.g., the case of Jack Mackerel in the South Pacific or Southern Bluefin Tuna in the Atlantic).

According to Brooks et al. (2014), amid declining stocks and an immediate need for quick action to reverse the downward trends, decision-making often becomes paralyzed as states compete to gain a portion of a diminishing catch quota.

Other reasons provided in Sumaila et al. (2007) for the ineffectiveness of RFMOs are as follows: RFMOs have limited powers to enforce their rules; the free-rider problem, i.e., states that choose not to join RFMOs continuing to fish outside of RFMO rules and thus undermining conservation measures; illegal, unreported, and unregulated (IUU) fishing is widespread on the high seas; massive subsidies are paid to the fishing sector in many countries, fueling fishing on the high seas; and it is incredibly expensive to monitor the currently existing wide array of diverse management strategies.

Beyond the crucial factors discussed by the above authors, some more factors and variables might overcome the lack of effectiveness, and that may add more elements to this reflection. In this research, the influence of knowledge and science will be observed and analyzed. It is well known and established in theory that a fundamental principle underlying modern fisheries management is that management decisions need to be based on the "best available scientific information." This principle is embedded in the UNCLOS which mandates that the determination of allowable catches and other conservation measures for living resources in the high seas are based on the "best scientific information available to the states concerned." The need for scientific advice as the basis for management decisions and the establishment of RFMOs was further affirmed in the United Nation Fish Stock Agreement (UNFSA). In practice, the advice of science is not very often accepted in political decisions.

Based on Haas (2004b), "Knowledge can speak volumes to power. Current research from comparative politics, IR, policy studies, and democratic theory suggests that science remains influential if its expertise and claims are developed behind a politically insulated wall. Epistemic communities are the transmission belts by which new knowledge is developed and transmitted to decision-makers."

According to this approach, "when usable knowledge is successfully constructed and transmitted, it yields distinctive results: regimes are developed by process of social learning; the regime rules reflect scientific consensus about environmental sustainability, and the regimes tend to be more effective" (Haas and Stevens 2011).

Lodge et al. (2007) agreed with this approach. They stated that fishery managers and marine resource users' lack of political will to implement management measures according to scientific advice and effectively enforce and comply with those management measures are leading agreements to fail or be less effective. They left behind to address what the institutional design is or how an epistemic community may help fill the gap between science and policy within the realm of the RFMOs.

It is recognized that several factors might improve agreement effectiveness: the willingness of key states to exercise leadership, the existence of strong international institutions, and the presence of strong epistemic communities or transnational science networks (Haas 2000).

This book focuses on the role and influence of science and epistemic communities in three specific agreements. Thus, through a combination of elite interviews (Annex V) and process tracing, three Regional Fisheries Management Organizations (RFMOs)—CCAMLR, ICCAT, and CCSBT[7]—are analyzed.

Those agreements were chosen as they were considered, according to Cullis-Suzuki and Pauly (2010), respectively, high, medium, and low in their biomass recovery performance. Thus, as a starting point, it seemed an excellent parameter to evaluate the role of science in those agreements to analyze if those scoring better on the biomass ranking were the same agreements where the advice of science was heeded.

It is worth noting that the biomass recovery was not necessarily due exclusively to the epistemic community or an institutional design that enabled science to speak truth to power. Many independent variables could have affected this dependent variable. However, it is the first step, and as causal inference thus becomes a *process* whereby each conclusion becomes the occasion for further research to refine and test it, this research is contributing to further the inferences about the effectiveness of agreements, in a way.

At this very moment, it is essential to clarify that this book does not intend to make comparative case studies as it recognizes that the agreements are structured and built in very different ways that could make the comparison between them impossible. However, there is the intention to look for empirical evidence that power is listening to science to make some decisions. Those in power to listen depend on diverse conditions that will be explored here.

For all of those reasons, this book will be divided into five main chapters. As it could be no different, the first chapter will explore the entire theoretical

---

[7] CCAMLR: Commission on the Conservation of Antarctic Marine Living Resources; ICCAT: International Commission for the Conservation of Atlantic Tunas; and CCSBT: Commission for the Conservation of Southern Bluefin Tuna.

framework—constructivism—that embraces the idea of social learning as one of the factors that could potentially enhance more cooperation and effectiveness, stating that under conditions of uncertainty—such as those associated with contemporary globalization and highly technical issues—the key is to design policy analytic processes that enable actors to learn about the world and about each other.

This approach, which has been well developed in various articles by Prof. Peter M. Haas, looks at the influence of groups of experts on the reformulation of policy outcomes and indicates how states and leaders may come to realize that new attitudes and political decision-making procedures are necessary to face some environmental problems.

The chapter shows that there has been an increasing recognition in recent years of the need for RFMOs to improve their performance because of demands contained within current international agreements aimed at better conservation and management of fishery resources. However, the RFMOs institutional design analyzed in this chapter does not fully allow scientists to produce usable knowledge, as defined by Clark and Majone (1985) and demonstrated empirically in the Haas and Stevens (2011) chapter. The science is not yet fully safeguarded enough from politics, which may, despite the quality of science produced, not be considered valid by many decision-makers.

The first chapter corroborates with Haas and Stevens (2011) that knowledge can speak volumes to power, and with this comes more effectiveness in terms of problem-solving. However, to make it work for fishery agreements, expertise and claims must be developed behind a politically insulated wall, a wall that is not yet working properly in every situation or when a decision is made.

Thus, the following chapters present three empirical case studies and, therefore, are fundamental to discuss the role of knowledge on the international fishery agreements cooperation. The empirical chapters contain a general history of each agreement, a description of its institutional design, and the specific case chosen within the whole agreement to explore our central question: when does power listen to science?

In the CCAMLR chapter, the epistemic community was identified under the Working Group of Fisheries Stock Assessment (WG-FSA). Furthermore, CCAMLR has been able to implement measures that serve as an example of how we might govern high seas resources more responsively and sustainably by always accepting scientific advice. They act on behalf of science. The chapter argues that no crisis or special event needs to trigger a change of procedure in this case. The epistemic community operates under its own institutional design.

For ICCAT, in the second chapter, the Bluefin Tuna case demonstrates that when power listens to science and implements scientific advice under a social learning process, fishery stocks can recover, thus reaching the primary goal of the Commission, conferring more effectiveness on the international agreement. However, in ICCAT, this only occurred after the widely publicized crisis about global tuna fishery management. In this case, an epistemic community emerged from the SCRS consisting of the ICCAT chairman, a few NGOs, several individuals, and some national scientists.

The Bluefin Tuna case under ICCAT illustrates that when decisions are based on the best scientific data available, and there is the political will to adopt the management measures required for the recovery of stocks, fisheries management works, even with a committee where the decision depends on the convergence of views from more than 50 countries.

On the other hand, the CCSBT presented a completely different situation. The countries were not relying on each other's scientific information. Each country's point of view almost always biased the knowledge. From the beginning, Japan, Australia, and New Zealand struggled to find the proper TAC that would keep them fishing without collapse or make the fish stock decline. They were successful only when a high level of independent scientists were able to help the process by creating an independent advisory panel and a management procedure sound enough to be credible by the decision-makers. The process, which took almost 5 years, made an insulated wall, as recommended by Haas and Stevens (2011), or Clark and Majone (1985), as a proper condition to promote the social learning process.

Finally, the conclusion reflects how the constructivist theory explains international cooperation and, therefore, creates some insights into the governance of the oceans. As mentioned previously, this reflection does not intend to compare the case studies between them due to the significant differences in what their institutions represent. It only highlights essential conclusions based on specific empirical data on the relationship between science and power. Yet, this does not mean to conclude that constructivism is the only theory that will explain all international cooperation cases on the environmental arena to the detriment of other theories. This book is guided by the empirical data, and for these case studies, constructivism supports the explanations of the specific cases analyzed here.

To conclude, this research contributes to knowledge on the sustainability science field on the complex and unsolved debate about how science can influence policy decisions, speaking truth to power, and why power must listen to science before making decisions on international agreements.

References

Barkin JS, Desombre ER (2013) Saving global fisheries: reducing fishing capacity to promote sustainability. MIT Press

Brooks CM, Weller AJB, Gjerde BK, Sumaila CUR, Ardron DJ, Ban ENC et al (2014) Challenging the right to fish in a fast-changing ocean. Stan Envtl LJ 33:289–457

Clark WC, Majone G (1985) The Critical appraisal of scientific inquiries with policy implications. Sci Technol Hum Values 10(3):6–19

Cullis-Suzuki S, Pauly D (2010) Failing the high seas: a global evaluation of regional fisheries management organizations. Mar Policy 34(5):1036–1042

de Bruyn P, Murua H, Aranda M (2013) The Precautionary approach to fisheries management: How this is taken into account by Tuna regional fisheries management organisations (RFMOs). Marine Policy 38:397–406

FAO (2001) International Plan of Action to prevent, deter and eliminate illegal, unreported and unregulated fishing. FAO, Rome, 24 p

FAO (2020) The state of world fisheries and aquaculture 2020. Sustainability in action. Rome. https://doi.org/10.4060/ca9229en

Gjerde KM, Currie D, Wowk K, Sack K (2013) Ocean in peril: reforming the management of global ocean living resources in areas beyond national jurisdiction. Mar Pollut Bull 74(2):540–551

Grotius H (1604) Mare liberum [the freedom of the seas] (Ralph Van Deman Magoffin trans, Oxford University Press 1916) (1608)

Haas PM (2000) Prospects for effective marine governance in the NW pacific region. Mar Policy 24(4):341–348. and "letter to the editor", Marine Policy, 24(6) 499–500

Haas PM (2004b) When does power listen to truth? a constructivist approach to the policy process. J Eur Publ Policy 11(4):569–592

Haas PM, Keohane RO, Levy MA (1995) The effectiveness of International Environmental Institutions. In: Haas PM, Keohane RO, Levy MA (eds) Institutions for the Earth: sources of effective international environmental protection. MIT Press, Cambridge, pp 3–24

Haas P, Stevens C (2011) Organized science, usable knowledge and multilateral environmental governance. In: Governing the air: the dynamics of science, policy, and citizen interaction, p 125

Lindley J, Techera EJ (2017) Overcoming complexity in illegal, unregulated and unreported fishing to achieve effective regulatory pluralism. Marine Policy 81:71–79

Lodge M (2007) Managing international fisheries: improving fisheries governance by strengthening regional fisheries management organizations. Energ Environ Develop Program 7:1–7

Lodge MW, Anderson D, Løbach T, Munro G, Sainsbury K, Willock A (2007) Recommended best practices for Regional Fisheries Management Organizations: report of an independent panel to develop a model for improved governance by Regional Fisheries Management Organizations. Chatham House, London. 141 p. http://www.oecd.org/dataoecd/2/33/39374297.pdfS

Miles EL, Arild U, Steinar A, Jørgen W, Jon Birger S, Carlin EM (2002) Environmental regime effectiveness: confronting theory with evidence. MIT Press, Cambridge, MA

Mitchell RB (2003) International environmental agreements: a survey of their features, formation, and effects. Annu Rev Environ Resour 28(1):429–461

Sumaila UR, Zeller D, Watson R, Alder J, Pauly D (2007) Potential costs and benefits of marine reserves in the high seas. Mar Ecol Prog Ser 345:305–310

Young OR (1999) The effectiveness of international environmental regimes: causal connections and behavioral mechanisms. MIT Press, Cambridge, MA

Contents

Acronyms

ATCPs	Antarctic Treaty Contracting Parties
ATS	Antarctic Treaty System
BIOMASS	Biological investigations on marine Antarctic systems and stocks
CCAMLR	Commission for the Conservation of Antarctic Marine Living Resources
CCSBT	Commission for the Conservation of Southern Bluefin Tuna
CDS	Catch documentation scheme
CITES	Convention of International Trade in Endangered Species
CPRs	Common pool resources
CPs	Contracting parties
CPUE	Catch per unit effort
CTCA	Commission for Technical Cooperation in Africa
DWFN	Distant Water Fishing Nation
EBFT	Eastern Atlantic and Mediterranean Bluefin Tuna
EC	European Commission
ECCSBT	Extended Commission of CCSBT
EEZ	Exclusive Economic Zone
ENGO	Environmental Non-Governmental Organization
Epicom	Epistemic Community(ies)
ERSWG	Ecologically Related Species Working Group (CCSBT)
ESC	Extended Scientific Commission of CCSBT
EU	European Union
FAO	United Nations Food and Agriculture Organization
ICCAT	International Convention for the Conservation of Atlantic Tunas
ICES	International Council for the Exploration of the Sea
ICNAF	International Commission for the Northwest Atlantic Fisheries
IR	International Relations
ITLOS	International Tribune for the Law of the Sea
IUU	Illegal, Unreported and Unregulated
MCS	Monitoring Control and Surveillance
MOU	Memorandum of Understanding

MP	Management Procedure (at CCSBT)
MSY	Maximum Sustainable Yield
NAFO	Northwest Atlantic Fisheries Organization
NEAFC	North East Atlantic Fisheries Commission
NGO	Non-Governmental Organization
OM	Operating Models
RFMO/A	Regional Fisheries Management Organization/Agreements
SAG	Stock Assessment Group (at CCSBT)
SBT	Southern Bluefin Tuna
SC	Scientific Committee (at CCSBT)
SCAR	Scientific Committee on Antarctic Research
SCOR	Scientific Committee on Oceanic Research
SCRS	Standing Committee on Research and Statistics
SSB	Spawning stock biomass
TAC	Total Allowable Catch
UNCLOS	United Nations Convention on the Law of the Sea
UNFSA	Agreement for the Implementation of the Provisions of the United Nations Convention on the Law of the Sea of 10 December 1982 Relating to the Conservation and Management of Straddling Fish Stocks and High Migratory Fish Stocks
WG-FSA	Working Group on Fisheries Stock Assessments (at CCAMLR)

List of Figures

List of Tables

Chapter 1
Regional Fisheries Management Organizations: Are They Effective?

Introduction

The agreement effectiveness study field is far from being fully explored. The study about which factors make international environmental cooperation more or less effective is likely to be one of the most critical aspects of research into the global environment's politics (Desombre 2007; Mitchell et al. 2020).

Why are some international environmental regimes more successful than others in solving problems that motivate their establishment? Why do some environmental regimes have a greater impact than others on the behavior of those actions that have given rise to the relevant problems? These are the questions that have been raised by many researchers dedicated to comprehending the international system where those agreements exist (Young 1999; Miles et al. 2002. Mitchell et al. 2006). Still, after 20 years, the agreement effectiveness field has kept its relevance, bringing more complexity evolving the architecture and agency concepts in the Anthropocene epoch (Biermann and Kim 2020; Hanusch and Biermann 2020).

One of the political science and international relations research areas that have addressed this effectiveness agreement discussion is the constructivist theory, which states that under conditions of uncertainty—such as those associated with contemporary globalization and highly technical issues—the key is to design policy-analytic processes that enable actors to learn about the world and each other. This approach, expanded in various articles by Prof. Peter M. Haas, looks at the influence of groups of experts on the reformulation of national or international objectives and indicates how states and leaders may realize that new attitudes and political decision-making procedures are necessary to face some environmental problems.

To maintain higher social learning within the agreement, the regime design must keep the construction of science insulated and build up clear procedures for the connection of science to policy decisions.

This connection is usually done by what Haas called the "transmission belt," formed by a group of scientists widely recognized as an epistemic community (Haas 1992; Cross 2013, 2015; Haas and Stevens 2011).

© The Author(s), under exclusive license to Springer Nature Switzerland AG 2021
L. R. Gonçalves, *Regional Fisheries Management Organizations*,
https://doi.org/10.1007/978-3-030-70362-2_1

In this context, this chapter will present the entire theoretical framework of constructivism and epistemic communities. It will empirically analyze the institutional design of three RFMOs that were considered, respectively, high, medium, and low in their biomass recovery performance: CCAMLR, ICCAT, and CCSBT[1] to evaluate if their institutional design is enabling science to influence policy outcomes and if the knowledge produced is organized in such a way as to change policy outcomes, thereby promoting more effectiveness to the agreements in terms of problem-solving.

Next, arguments will be presented to show that effectiveness depends on diverse factors. Institutional scientific design plays an important role in enabling experts to produce usable knowledge to influence decision-makers to make the agreements more effective.

This chapter aims to establish a framework of analysis and examine the institutional design for science to understand how it affects or influences the agreement's performance.

Constructivism as a Theoretical Approach

According to Adler (2002), all constructivists (modernist, modernist linguistic, and critical)—with the exception, perhaps, of the extreme postmodernist wing of radical constructivism—share two concepts: what Stefano Guzzini (2000) summarized as the social construction of knowledge, and the construction of social reality. In combination, these concepts are constructivism's common ground, the view that the material world does not come categorized. Therefore, the objects of our knowledge are not independent of our interpretations and our language. This means that different collective meanings are attached to the material world twice, once as social reality and again as scientific knowledge. In other words, knowledge is both a resource that people use in their day-to-day life for the construction of social reality and the theories, concepts, meanings, and symbols that scientists use to interpret social reality.

This dichotomous description is offered for analytical purposes only. For reflexive knowledge or an interpretation of the world when imposed on material reality, becomes knowledge for the world—the power to change the world by collective understandings and, concurrently, with human motives and intentional acts. The above analysis means not only that there is no perfect correlation between objects "out there" in nature and our classifications of nature, but also that social facts, which are the objects of its study, emerge from the interaction between knowledge and the material world, neither of which is invariant.

---

[1] CCAMLR: Commission on the Conservation of Antarctic Marine Living Resources; ICCAT: International Commission for the Conservation of Atlantic Tunas; and CCSBT: Commission for the Conservation of Southern Bluefin Tuna.

Unlike rationalism in both major schools—neo-realism and liberal institutionalism—which take the world as it is, according to Adler (2002), constructivism sees the world as a project under construction, as becoming rather than being. Unlike idealism, post-structuralism, postmodernism, and radical constructivism, which take the world only as it can be imagined or talked about; constructivism accepts that not all statements have the same epistemic value and consequently some foundation for knowledge.

Emmanuel Adler, in his chapter "Constructivism in International Relations" (Adler 2002), states very clearly that constructivism, which reached the shores of IR in the 1980s, describes the dynamic, contingent, and culturally based condition of the social world. It has major implications for understanding knowledge, including scientific knowledge, and how to achieve it. Constructivism thus has the potential to transform the understanding of social reality in the social sciences.

Unlike other contemporary approaches like neo-realism, where the most powerful states have a decisive influence over the formation, nature, and effectiveness of a regime; and from liberal institutionalism, where international institutions themselves can bring about cooperation through creating an expectation of future gain, exchanging information, and building trust; the constructivist approach considers that governments can learn to apply new patterns of reasoning to the formulation of policy, which is reflected in a more sophisticated understanding of the complex array of causal interconnections between human environmental and economic activities (Haas 1990).

Constructivists focus on the role of ideas, norms, knowledge, culture, and argument in politics, stressing, in particular, the role of collectively held understandings on social life. One of the main characteristics that differ constructivism from realism and institutionalism is that human interaction is shaped primarily by ideational factors, not simply material ones (Finnemore and Sikkink 2001).[2]

Under the constructivism theory, Haas' works encourage the social learning approach. In the early 1980s, Ernst Haas (1983), Peter's father, suggested a powerful sociological role of international cooperation based on learning, that is, on the introduction to politics of scientific consensual concepts. Borrowing the concept of *episteme* from Foucault, Ruggie (1975) further developed this program, which Haas (1992), and Adler and Haas (1992), turned into an agent-oriented constructivist research program on "epistemic communities."

Over the past two decades, this research pathway has revealed the existence of numerous highly influential epistemic communities—basically defined as expert networks—that have been involved in the trajectory of international cooperation by

---

[2] The constructivist theory was described as detailed by Adler (2002) in the fifth chapter, and it was used as a reference for this research. It is very important reading for those who want to understand this theory's roots, which I will not describe in detail in this book. In the same book, Fearon and Wendt (2002) wrote a chapter entitled "Rationalism v. Constructivism: A Skeptical View," and it is worth it to read to understand that within IR there is no consensus under the battle of theories. In this chapter, they claimed, "constructivism is not a 'theory' at all, any more than is rationalism" (p. 74).

virtue of their shared professional expertise. It has become a progressive research program, and since then, the concept was applied to numerous examples (see the International Organization Special Edition of 1992 for the main reference). Twenty years later, the approach has been revisited by Cross (2013), and even further discussed by Peter Haas throughout his career, making him the main reference for this particular research program (Haas 2014, 2015).

The study of epistemic communities forms part of the constructivist theory; it explains sources of actors' understandings in a complex and uncertain policy environment and the actor's attendant behaviors or practices under specified conditions.

Besides all of the efforts, diverse researchers have done to further the inquiry using the epistemic communities concept on many subjects, including European security (Cross 2015), whaling (Peterson 1992), the Mediterranean (Haas 1990), and biodiversity (Inoue 2003); however, international fisheries have not yet been explored under this framework.

The Science and Epistemic Communities on International Environmental Agreements

Despite widespread agreement on the importance of science for policy-making, there are still divergent perceptions of how, when, and under what conditions science influences policy, and accordingly, on how the interplay between science and policy should be best organized (Lidskog and Sundqvist 2015). Additionally, it is known that experts' involvement is found at all levels of government—international, national, and local and also with a range of policy actors (e.g., sectorial groups, citizens, and private entrepreneurship) (Dunlop 2016).

Comparative studies of environmental regimes confirmed that organized scientific expertise had a distinct influence over the effectiveness of environmental regimes in which they played a negotiating role (Andresen 2000; Haas 2007).

One approach for this analytical problem may be descriptive, addressing the variety of arrangements that can provide scientific expertise to international agreements (Watson 2005). Another approach uses the critical point of view for claiming a policy process democratically constructed that would involve not only elite scientists but also a larger number of stakeholders contributing deliberatively to the policy construction characterized as a process of co-production (Jasanoff 2013; Jasanoff and Martello 2004). Additionally, an approach developed by Peter Haas that will be used here addresses the concept of social learning and discusses how science and knowledge can or cannot contribute to more sophisticated management of technical issues (Haas 2004a, b).

The approach suggested and developed by Peter M. Haas argues that science can play an important role in shaping policy decisions through the development of an epistemic community. He emphasizes the importance of science, and in particular consensus-based knowledge in policy-making. Haas argues that consensus-based science can play an independent and important role by influencing and even refor- mulating state interests, thereby bringing about international agreements that tran- scend and reshape state interests. This is made possible through the involvement of experts. Thus, from the epistemic community perspective, environmental regimes are driven not only by state powers but also by epistemic networks under certain conditions.

The epistemic community argument was that, normatively, epistemic communi- ties ultimately provided better advice than other modes of policy advice because expert advice is likely to be warranted (Haas 2015).

Unlike other organized interest groups active in politics and policy-making, epis- temic communities have internal beliefs that make them more likely to provide information that is politically untainted and therefore more likely to "work" in the political sense that this information will be embraced and followed by political authorities concerned about the need to be impartial (Haas 2001, 2007, 2012, 2014, 2015; Haas and Stevens 2011).

According to Haas (2004, 2015), after a widely publicized shock or crisis, the states recognize the need to deal with the problem and then rely on scientists for help. Here it is all about the decision-makers' recognition of the limits of their abili- ties to deal with new issues and the need to defer or delegate to authoritative actors with a reputation for expertise.

In this context, when the experts are able to develop usable knowledge, the decision-makers feel compelled to apply scientific consensus, social learning emerge (Haas and Haas 2002). Thus, in this situation, expert networks are likely to emerge and can be crucial in shaping policy outcomes (Haas 1992, 2014; Lidskog and Sundqvist 2015).

The epistemic communities concept coined by Peter Haas as a professional net- work with authoritative and policy-relevant expertise was widely presented in Special Issue of International Organization entitled, "Knowledge, Power, and International Policy Coordination (1992)." In almost 30 years, the idea has gained some robustness in International Relations (IR) scholarship and applied as a frame- work for different subjects (Cross 2013, 2015; Haas 2015).

Epistemic communities are one of the main actors responsible for aggregating and articulating knowledge in terms of states' interests for decision-makers and dis- seminating those beliefs internationally. In a broader political context, epistemic communities provide one of the major channels by which overarching regime prin- ciples, norms, and rules are articulated for the international community and dis- seminated internationally. While epistemic communities are key agents responsible for disseminating principles, norms, and rules, the level of diffusion will largely depend on the political influence of its members: their ability to persuade others, their ability to consolidate bureaucratic influence in important institutional venues, and their ability to retain influence over time. Thus, states' interests and decisions to deploy state power are identified as subject to consensual knowledge (Haas 2014).

It is a concept invoked by constructivist scholars of IR to focus analytic attention on how states formulate interests and reconcile differences of interest. Epistemic communities are a principal channel by which consensual knowledge about causal understandings is applied to international policy coordination and by which states may learn through international cooperation processes (Haas 2012).

They are networks of knowledge-based communities with an authoritative claim to policy-relevant knowledge within their domains of expertise. Their members share knowledge about the causation of social or physical phenomena in an area for which they have a reputation for competence and a common set of normative beliefs about what actions will benefit human welfare in such a domain. In particular, they are a group of professionals, often from several different disciplines, who share the following essential characteristics (Haas 2012):

1. Shared principled beliefs. Such beliefs provide a value-based rationale for social action by the members of the community.
2. Shared causal beliefs or professional judgment. Such beliefs provide analytic reasons and explanations of behavior, offering causal explanations for the multiple linkages among possible policy actions and desired outcomes.
3. Common notions of validity: intersubjective, internally defined criteria for validating knowledge. These allow community members to differentiate between warranted and unwarranted claims about states of the world and policies to change those states.
4. A common policy enterprise: A set of practices associated with a central set of problems that have to be tackled, presumably, out of a conviction that human welfare will be enhanced as a consequence.

For this book, the epistemic community was identified through cross-checking attendance lists (Commission and Scientific Committee), semi-structured interviews with key stakeholders and consultations over time from the secondary and scientific literature.

Agreement Effectiveness: Challenges and Opportunities

The debate about whether "institutions matter" has no longer been necessary, as it has fallen into common sense. Saying that institutions matter implies that, due to institutions' existence, actors behave differently than they would in the absence of institutions or the presence of different institutions (Voigt 2013). International organizations and regimes are established to perform a particular function or achieve a specific goal. Thus, one of the fundamental questions to be asked about these institutions is how effective they are in delivering what they were established and designed to achieve (Hovi et al. 2003). Here is where the challenges lie.

The discussion about international environmental cooperation's effectiveness is challenging for several reasons (Desombre 2007). One of the reasons is that it is very complex to define what is meant by effectiveness and the factors best qualified to measure it. As for Young and Levy (1999), "regimes can range along a continuum from ineffectual arrangements, which wind up as dead letters, to highly effective arrangements, which produce quick and decisive solutions to the problems at hand."

As Desombre (2007) stated, sometimes these agreements have been effective at changing behavior[3] and ultimately beneficially impacting the environment. In other cases, success—either in the creation of international mechanisms or in their influence on the state or individuals' behavior, and ultimately their ability to improve the conditions of their environment—is less specific. Nonetheless, through these experiences, we can find lessons to learn to make international cooperation more effective and, more importantly, how to understand why it is effective.

The question about "why some international environmental agreements are more effective than others?" has been posed by other researchers such as Underdal (1992) and Young (1999), and it has always been a challenging discussion.

First, it is complex to define a concept for effectiveness in a particular context. Borrowing the definition from Young and Levy (1999), "effectiveness is a matter of the contributions that institutions make to solving the problems that motivate actors to invest the time and energy to create them." In this case, effective agreements would be those that improve environmental quality. But even this definition can be very ambiguous as an improvement in the environment could be directly related to the international agreements, or it could be a coincidence, or even a natural fluctuation that has nothing to do with human action (Desombre 2007).

Secondly, in terms of lessons learned, the critical thing to understand is why some agreements are more effective than others. What are the factors that are supporting effectiveness in those agreements? How can we measure effectiveness? The important thing is to develop an operational procedure for measuring effectiveness. Many authors are working to clarify and reflect on these methods. One of the broadest initiatives is the Oslo-Potsdam project (Hovi et al. 2003) and other important references, including Helm and Sprinz's (2000) paper, Miles et al. (2002), and Young's (1999) books.

According to Mitchell (2003), the agreement's effectiveness can vary depending on member countries' characteristics, the international context, and the underlying environmental problem as to the differences in agreement design.

As you can see, there are many approaches that one could take to study agreement effectiveness. For Young and Levy (1999), it may be a political, normative, economic, legal, and problem-solving approach. Each of them presents the pros and cons, and they may be used analytically, separated, or in groups, depending on the objective of the research.

For this research, effective arrangements entail policy changes by states according to the intentions of negotiated treaties that lead to, or are likely to lead to, improvements in environmental quality. Effectiveness means that the knowledge input from the epistemic communities induces states to change their behavior in ways that promoted the achievement of negotiated aspirations, especially the improved biomass of each stock studied here.

---

[3] See Saving the Mediterranean: The politics of international environmental cooperation (Haas 1992) and Banning chlorofluorocarbons: epistemic community efforts to 187 protect stratospheric ozone (Haas 1992a)

In general, most people want to know if the agreements are improving environmental quality, as stated above. However, as the authors pointed out, problem-solving approaches present practical problems that are sometimes severe because the natural system is complex. It is very hard to assume that the agreements' improvement was caused by the agreements per se unless you use a counterfactual analysis.[4] For the fishery agreement's evaluation, the quantitative counterfactual analysis is even more difficult because most of the statistical data that could lead us to a debate about "problem-solving" is supplied by states. The information is neither accurate nor independent.

Thus, recognizing the challenges of measuring an agreement's effectiveness, and at the same time trying to avoid the discussion about how to measure an agreement's effectiveness that have been addressed by other researchers, this book has chosen the Cullis-Suzuki and Pauly (2010) paper that has quantitatively assessed the effectiveness of the world's 18 RFMOs and concluded that they have all failed, but on different levels. From here forward, what will be discussed is why the agreements have failed? Why did they fail on different levels?

Aiming to discuss the matter in this sense, the research will follow the process of tracing methodology for all three individual agreements, which involves "attempts to identify the intervening causal process—the causal chain and causal mechanism—between an independent variable (or variables) and the outcome of the dependent variable" (George and Bennett 2005).

There are 18 existing and prospective RFMOs with mandates to establish fishery conservation and management measures, which means that almost all of the global high seas are now covered by at least one RFMO. However, the effectiveness of current RFMOs has rarely been comprehensively assessed, despite indications of the decline of many high seas fish stocks (Myers and Worm 2003).

The Cullis-Suzuki and Pauly (2010) paper addresses this topic quantitatively for all RFMOs. In their work, global evaluation on the effectiveness of RFMOs is based on a two-tiered system: (1) in theory (or "on paper"), i.e., how well RFMOs meet standards as set by Lodge et al. and as measured by the comprehensiveness of available information; and (2) in practice, i.e., how well the stocks under RFMO management do, as measured by current abundance (biomass) trends of managed stocks, and supported by trends through time.

In general, the RFMOs have as an objective to "establish fisheries conservation and management measures," thus, the current abundance (biomass) seems a good criterion to measure an agreement's effectiveness in terms of problem-solving.

With that in mind, the ranking they produced in their paper scores (*Q scores*): CCAMLR (100%), ICCAT (37.5%), and CCSBT (0.0%), are best, medium, and low performances. Even so, in the end, they recognize that "it is evident from the results here that the priority of RFMOs—or at least of their member countries—has been first and foremost to guide the exploitation of fish stocks. While conservation

---

[4] To understand how counterfactual can be applied to measure effectiveness, read: Oslo-Potsdam Solution Project Hovi et al. (2004), and also have a look at Breitmeier et al. (2011).

is part of nearly all of their mandates, they have yet to demonstrate a genuine commitment to it on the water." Furthermore, stating that they concluded the paper by questioning, "Why have RFMOs failed?"

Usable Knowledge and Epistemic Communities

As explained above, diverse authors have discussed the factors that can ensure more effectiveness for the agreements (Young and Levy 1999; Miles et al. 2002). Some say that "ensuring that agreements contain provisions that are responsive to the type of problem being addressed" (Mitchell 2003), and others prefer to elaborate a list of factors as Ridgeway (2014).

Peter Haas, Robert Keohane, and Marc Levy suggest that effective regimes are built upon existing concerns, work with or create capacity, and occur in an appropriate contractual environment (Haas et al. 1995).

Haas (1992) also argues that the diffusion of new ideas and information can lead to new patterns of behavior, which proves to be an essential determinant of international policy cooperation.

However, following the ideas of Haas and Stevens (2011), knowledge operates and changes behavior when organized and transmitted in a way that policy-makers can understand and trust. In this context, regimes developed by social learning and whose rules reflect scientific consensus about environmental sustainability tend to be more effective. And that is what this research seeks to evaluate the three specific RFMOs.

One of the major research areas developed by Peter Haas has been the influence of science and usable knowledge in the outcomes of the international agreement, usually promoted by the epistemic communities—knowledge-based groups of experts and specialists who share common beliefs about cause-and-effect relationships in the world and some political values concerning the ends to which policies should be addressed.[5]

When the decision-makers feel they need the information to make a decision and lack it, epistemic communities are one possible provider of this sort of information and advice. As demands for such information arise, networks, or communities of specialists capable of producing and providing the information emerge and proliferate (Haas 1992).

Peter Haas started working on epistemic communities when he was doing his dissertation on the Mediterranean Sea (1982–1983) (Haas 1989, 1990). Through elite interviews with many members involved in the Mediterranean pollution debate, he noticed that many of them presented shared beliefs and a consensus of knowledge that influenced their policy decisions.

---

[5]A special edition of the International Organization journal was published about epistemic communities for diverse subjects in 1992. See references for complete information.

For Haas (2014), "epistemic communities analysis provided a delegation model in which decision-makers construct their political realities based on the technical advice provided by experts. Intended effects by one set of actors (agents) lead to unintended effects by other actors (principals) with aggregate social benefits through the provision of international public goods from focused collective action."

This provides an excellent framework of analysis to discuss the fishery agreements' effectiveness once we face agreements that operate on the edge of uncertainty, where many states and non-state actors are performing diverse roles with no consensus in most decisions.

Subscribing to the definition of epistemic communities from Haas (2014) as "epistemic communities is a concept applied by constructivist scholars of political science to focus analytic attention on the process by which states and other political actors formulate their interests and reconcile differences of interest. Epistemic communities are a principal channel by which consensual knowledge about causal understandings is applied to international policy coordination, and by which states may learn through international cooperation processes."

Under conditions of uncertainty, decision-makers have various incentives and reasons for consulting epistemic communities, some of them are more politically motivated than others (Haas 1992). These new actors could drive governments to recognize and follow new interests in environmental protection so that they were willing to resist systemic forces that would push them to pursue more constrained and transitory agreements.

It is rare to see science influencing policy. Besides all the knowledge that has been generated by scientists to offer solutions to solve environmental problems, it is crucial to recognize that not all of it has been absorbed by the decision-makers (Haas and Stevens 2011), and generally a "problem of fit still persist" (Young 2003). Also, it is not possible to neglect the conventional approach that stresses the role of interstate power. However, the discussion here is the institutional design that makes experts provide usable knowledge to the decision-makers.

Consider also that politics do not always accept science as a universal truth. Sometimes science can be guided, not always by independent scientific principle, but by the sponsors' agenda or by the "politicization of science" (Pielke2004) so that science may unconsciously reflect such hidden values (Haas 2004a, b) and that all those factors, and others not listed here, could potentially lead decision-makers to distrust science even when it is in their favor.

However, there are good examples of where power has listened to science. For instance, the stratospheric ozone and European acid rain efforts are widely hailed as among the more successful and effective international environmental governance efforts of the contemporary era (Haas 2001; Miles et al. 2002). Or, consider the Mediterranean plan to cope with pollution (Haas 1989). On the other hand, there are anomalous cases where the argument failed, such as desertification and whaling (Haas 2015).

It seems that decision-makers are likely to adopt science when the institutional design of the agreement enables an organized scientific view, and it includes an insulated and robust group of experts holding a "usable knowledge."

Although usable knowledge has been used in different contexts and situations, in short, usable knowledge encompasses a substantive core that makes it usable for policy-makers and a procedural dimension that provides a mechanism for its transmission from the scientific community to the policy world (Haas and Stevens 2011).

Clark and Majone (1985) made an interesting reflection on the relationship between scientists and policy-makers and why they are sometimes dissatisfied with each other. To produce usable scientific knowledge, they defined some important criteria such as adequacy, value, legitimacy, and effectiveness: adequacy relates to including all of the relevant knowledge or facts; value has to do with contributing to further understanding and meaningful policy; legitimacy relates to the acceptance of the knowledge by others outside of the community that developed it; effectiveness relates to its ability to shape the agenda or advance the state of the debate and ultimately to improve the quality of the environment. The fact is that a scientist does not have all of this at hand at all times.

Constructivist approaches to policy analysis suggest that science must be developed authoritatively and then delivered by responsible carriers to politicians. "The transmission belt of like-minded scientists is called an 'epistemic community,' and the more autonomous and independent science is from policy, the greater its potential influence" (Haas 2001, 2007).

In this context, regimes built that have decisions based on usable knowledge produced by independent science appear to be more effective at inducing states to achieve their intended goals of improving environmental quality (Haas and Stevens 2011).

Fisheries Policies, Science, and Usable Knowledge

When scientific knowledge comes to fishery and ocean affairs, it seems to have a history of this practice, starting with the mandate in Article 61 of the United Nations Convention on the Law of the Sea (UNCLOS) for decisions to take into account "the best scientific information available." The need for scientific advice as to the basis for management decisions and the establishment of RFMOs was further affirmed in the United Nation Fish Stock Agreement (UNFSA).[6]

Science is invoked to convey messages that decisions made have taken into account all of the relevant information, processed the information systematically and soundly dealt with the information in objective, verifiable and balanced ways, leading to rational decisions given the information available (Pielke2004). However, "science-based" does not mean that all of the sources and types of information were

---

[6] United Nations Agreement for the implementation of the provisions of the United Nations Convention on the Law of the Sea of the tenth of December 1982 relating to the conservation and management of straddling fish stocks and highly migratory fish stocks, 1995. http://www.un.org/depts/los/fish_stocks_conference/fish_stocks_conference.htm#Agreement.

provided equal weight in the decisions, nor that "scientific information" comprised the only factor considered in the decision-making (Mitchell et al. 2006).

Garcia et al. (2014) created an incredibly relevant work updating their past papers from 2012 on fishing and biodiversity management. It shows the involvement of science in different phases of the process.

According to Garcia et al. (2014), from 1850 to 1900, science occupied a place on the fishery policy discussion, but it is still too early to say if that knowledge was influencing politics.

From 1900 until today, there were many steps where science was growing and developing more information on fisheries while seeking to subsidize policy outcomes. The creation of ICES[7] has helped significantly in bringing science into policy-making decisions. The complete history is from Garcia et al. (2014). However, the facts show that science has been present throughout most fishery development periods but has not always influenced it based on acquired knowledge.

Mora et al. (2009) studied the effectiveness of fishery management regimes worldwide and calculated the probable sustainability of reported catches to determine how management affects fishery sustainability when scientific knowledge is applied. Their results claimed that the conversion of scientific advice into policy, through a participatory and transparent process, is at the core of achieving fishery sustainability. However, only a few countries have a robust scientific basis for management recommendations and transparent and participatory approaches to convert those recommendations into policy while ensuring compliance with regulations. So, uncertainty also plays an important role in this fishery scenario.

States also have distinct views on how environmental issues mainstream into agendas that matter to them. These diverse interests can be challenging to resolve, which can reasonably be achieved in global decision-making.

In this light, Haas (2004a) observes that scientific assessment bodies' design is critical to decision-making processes, including the need for resulting information to be understandable to decision-makers. His criteria for science legitimacy for institutional decision-making are accuracy (widely perceived to be true), legitimacy (achieved through impartial processes insulated from direct political influence), and saliency (policy-relevant and politically tractable). Science is less successful in anchoring solutions when there is: public suspicion over its development or methodology; institutional design questions; lack of clarity on priorities for scientific assessment; and governmental misgiving or unwillingness to cede "authority" to science resulting in policy decisions that have unfavorable consequences.

The design of the knowledge-creating processes and access to their outputs is vital to the utility and credibility of governance, as Ridgeway (2014) pointed out. With organized scientific input to the policy process, negotiated outcomes are much

---

[7] ICES—The International Council for the Exploration of the Sea (ICES) is a global organization that develops science and advice to support the oceans' sustainable use.

ICES is a network of more than 4000 scientists from over 350 marine institutes in 20 member countries and beyond. One thousand six hundred scientists participate in our activities annually. (http://www.ices.dk/Pages/default.aspx).

more likely to yield integrated management efforts rather than mere political compromises (Haas 2006).

The nature and value of the scientific advice provided to RFMOs are shaped by independent variables that include the institutional and operational arrangements established by the RFMO, how the advice is framed, and the quality and timeliness of data underlying the advice (Willock and Lack 2006).

The three selected agreements have their institutional design for scientific influence built in as described in Table 1.1 (adapted and updated from Haas and Stevens 2011).

From there, it is clear to note that the institutional design for these agreements is very similar. In general, they have failed to insulate scientific bodies from direct or indirect political control. The design of the scientific committees in fisheries reflects

Table 1.1 Information about the institutional design of science for the agreements

	CCAMLR	ICCAT	CCSBT
Convention objective	The conservation of Antarctic marine living resources (for this convention, the term "conservation" includes rational use)	Maintaining the populations of these fishes at levels which will permit the maximum sustainable catch for food and other purposes	To ensure, through appropriate management, the conservation and optimum utilization of Southern Bluefin tuna
Committee organization	Standing; sets agenda	Ad hoc	Standing
Expert selection	States	States	States
Type of science committee	Open	Open	Open
Decisions are made by	Consensus	Majority	Consensus
Decisions (binding/ advisory)	Binding	Binding	Binding
Objection to decisions	No	Yes	No
Degree of social learning	Social learning—It happens through its own institutional design	Little/none—It may happen but it depends a lot from the involvement of other stakeholders to create an environment where knowledge is warranted	Little/none—It has been improving with time and with the management procedure approval. It seems that decision-makers are now trusting in the knowledge generated by the independent advisors
Data provided by	Government	Government	Government

an institutional pathology that limits science's ability to speak to power, however, on different levels and in different situations.[8]

The CCAMLR website shows that they give importance to science when they state "science is fundamental to CCAMLR," and also add that the Convention requires the Commission to take "full account of the recommendations and advice of the Scientific Committee and this emphasis on science was reiterated in 2009 in Resolution 31/XXVIII[9] on the use of the best available science."[10]

It is not for nothing that CCAMLR has been considered as one of the most successful RFMOs and it scored better, according to Cullis-Suzuki and Pauly (2010), in terms of biomass recovery (dependent variable).

Haas (2006) and other researchers (Constable 2011; Brooks 2013; Brooks et al. 2014) also corroborate that CCAMLR is a good example of effective agreement; fish stocks are recovering, and standards focused on environmental issues have been enhanced, adopted, and implemented. As pointed out by Brooks et al. (2014), one reason for CCAMLR's success in passing more ecosystem-based and conservation measures is that it is comprised of both fishing and non-fishing science-focused members.

However, Constable (2011) noted that, although CCAMLR has been doing well in its recovery of the fish population, science has not yet been considered fully part of the work of CCAMLR. Attention needs to be given to developing the capability and tools to help overcome differences amongst scientists in providing managers advice. As Constable remarks, "One of the greatest impediments for CCAMLR is achieving consensus over science before agreeing by consensus to the management actions that are dependent on that science."

In other words, this means that there is an institutional design promoting the flow of science for decision-makers in their organogram where the Scientific Committee lies right below the Commission itself. However due to the lack of agreement between the scientists, there is no consensus on information, which means that there is a lack of usable knowledge to provide an even higher social learning experience.

Nonetheless, when CCAMLR is compared to ICCAT and CCSBT, even with those criticisms and the need for improvements, CCAMLR's is the agreement that presents more social learning and science-based policy than the others, and by consequence, the one that is more effective in terms of problem-solving. This corroborates with Cullis-Suzuki and Pauly (2010), where CCAMLR scored (Q score) 100% in its performance. Even considering that CCAMLR is not typical, it would be illuminating towards reviewing other institutional characteristics that might facilitate more RFMOs (Brooks et al. 2014). Still, on the science side, it seems there remain many improvements to be made.

---

[8] The next chapters will show specific case studies to illustrate situations where it occurs and when science speaks to power under certain conditions.

[9] https://www.ccamlr.org/en/resolution-31/xxviii-2009.

[10] Website—https://www.ccamlr.org/en/science/science. Accessed in: the tenth of December 2014.

ICCAT and CCSBT, in turn, were placed with medium- and low-performance scores, respectively, in Cullis-Suzuki and Pauly's (2010) paper. Haas and Stevens (2011) considered that they present little or no degree of social learning, which is reflected in their biomass recovery performance.

ICCAT has a history of not listening to science (KVIST, no date). Eastern Bluefin Tuna (EBFT) can help illustrate the ICCAT's low level of social learning as in the past; they have consistently ignored their scientists' advice regarding this species. At first, the Commission chose to adopt regulations without enforcing them, then in 1974 ICCAT passed a binding recommendation to limit harvests to "current levels" as per scientific advice. The Commission frequently set legal catch levels higher than those recommended by the SCRS but quite close to the estimated total harvest. Although SCRS advice became more optimistic later that decade, the gap between legal catches and scientific advice remained quite large, as illustrated by Webster (2011). In 2011/2013 this changed when the Commission, for the first time, accepted the Scientific Committee suggestion quota for EBFT. The decision was welcomed even by international NGOs who often complained of the lack of science in ICCAT.[11] This case will be explained in detail in Chap. 3.

However, as highlighted by Aranda et al. (2010) in their report review on RFMOs, research and assessment of each RFMO depend on its science structure. In ICCAT's case, the scientific input comes from its working groups, composed of Member States scientists. Data for the scientific process is generally supplied by the Member States, including total catch, catch and efforts data, and catch and size data. However, data submission is often incomplete or late from states and may even be underreported, which may jeopardize data used in sound management advice. Also, they added that detailed operational data is rarely supplied by members and is considered highly confidential.

For CCSBT, the Scientific Committee considers stock assessment analyses conducted by national scientists and consultants and is aided in this by the independent Chair of the Scientific Committee and an independent expert panel. This panel's role is to facilitate consensus and, if this proves to be impossible, to provide its independent view to the Commission (Aranda et al. 2010).

Apart from the necessary progress CCSBT has made by including this independent expert panel, Polacheck (2012) published a paper with a case study that shows one occasion, perhaps one of many, where a scientific paper was presented, discussed, and used in the formation of a decision at the 2006 meeting of the Commission for the Conservation of Southern Bluefin Tuna (CCSBT), the Stock Assessment Group (SAG), and the Scientific Committee (SC), which was summarized in the reports from those meetings but then subsequently withdrawn as a consequence of international and domestic political concerns rather than scientific issues.

---

[11] http://blogs.nature.com/news/2012/11/conservationists-claim-victory-for-science-over-tuna-quotas.html and http://mediterranean.panda.org/?206761/Decisions-on-Eastern-Atlantic-and-Mediterranean-bluefin-tuna-follows-scientific-advice.

According to Polacheck (2012), examples of this issue occurring by several nations were observed during 20 years of personal involvement with the scientific processes of RFMOs and from his personal experience.

The provision of independent scientific advice is central to the operation of RFMOs even though RFMOs are largely policy (and political) instruments. Political pressures and intervention in science and the crossing of the boundaries between scientific and political processes appear to be relatively common (Polacheck 2012).

According to Haas and Stevens (2011, pp. 139–140), there is a criteria list, based on previous publications (e.g., Haas 2004b), that would discuss an agreement's effectiveness by looking at the institutional scientific design.

Most social learning treaties have standing environmental monitoring and research committees to provide timely warnings of new problems, monitor achievements of regime goals, and educate politicians and policy-makers on environmental issues (Haas and Stevens 2011). But this is not the case of the fishery agreements studied here, as all the scientific information is provided by the governmental bodies that are not always able to provide accurate information, nor are they able to continuously supply the information. By evaluating their performance reviews (CCAMLR, ICCAT and CCSBT 2008), it is possible to see that CCAMLR has less dependence on government information than the others.

The maintenance and support of scientific bodies within multilateral environmental governance arrangements are vital for constructing usable knowledge within the regime. Standing scientific panels allow the constant construction and transmission of accurate and timely information. Three types of standing-committee structures are typical in multilateral environmental regimes (Haas and Stevens 2011).

The first type includes scientific groups that set their own schedules and research agendas. These groups meet as they see fit between meetings of policy-makers and according to their own determined needs. The CCAMLR Scientific Committee represents this first type. The Scientific Committee meets annually and immediately prior to the Commission meeting. To address the wide range of scientific areas that might impact the decisions of the Commission, the Scientific Committee has established many working groups that meet during the year; they establish their agenda and assist in formulating scientific advice on key areas.

The second type is made up of those groups whose meeting the COP sets times. For example, most fishery agreements require their scientific bodies to meet a month or 2 months prior to the COP meeting, like ICCAT and CCSBT, for instance.[12]

Finally, the third type includes ad hoc science panels, which, especially if called for by the Member States, introduce a high level of political involvement in science reporting, which also matches with ICCAT's design.

In addition, long intervals between reports impede the timely involvement of science in policy discussions. Standing committees that set their agenda and schedule appear to be the most politically insulated, whereas ad hoc science panels appear to be the most vulnerable to political involvement.

---

[12] http://www.iccat.int/en/meetings2015.htm.

The CCSBT has made progress in protecting science from policy by establishing an Advisory Panel to provide external input to its stock assessment and scientific processes. It has also appointed an independent chairperson for the Scientific Committee who does not represent any government view.

Additionally, another factor that may be evaluated is the choice of scientists. According to Haas and Stevens (2011), the most successful agreements have a selection of scientists from secretariats or even scientific bodies of other intergovernmental organizations. That is not the case for any of the agreements discussed here. All scientists under the Scientific Committee are designated by the countries themselves, which does not ensure the safeguarding of scientific information.

In these selected fishery agreements, the design of a scientific body is open to all of them, which allows each member country to appoint a representative to a scientific body. However, this does not typically create proper safeguards between the scientific bodies and the policy-makers.

From this chapter, it is possible to learn that CCAMLR has the best institutional design for science. It allows science and the epistemic community to produce and influence the decisions with the addition of their ideas and knowledge. Some of these characteristics could even be shared with other agreements to enhance best practices for science's influence on policy decisions. However, it seems that the scientific community is not speaking in one unique voice yet, and this dissonance and lack of consensus is circumventing an even higher level of effectiveness. The lack of consensus promotes uncertainty, which is one of the significant reasons why long-term actions such as biodiversity protection or fish stock rebuilding are difficult to implement.

According to the ICCAT performance review (ICCAT 2008a, b), while modeling and stock assessment is not a perfect science, the fisheries that are managed by ICCAT are reasonably well understood, and the SCRS is well-regarded and professional scientific body. However, lack of participation and lack of data provided by countries can be serious impediments to the work of SCRS and the Commission.

Nevertheless, ICCAT is not a good example of where science can be considered to be well-regarded. Some ICCAT scientists believe that they are independent of policy decisions; in most cases, participation in the scientific process revolves around member delegations with official heads of delegations. The heads of state delegation are most commonly government representatives.

In most cases, a large fraction of a member's national delegation is drawn from government fishery departments or associated marine research institutes, with a substantial portion of the funding for participation in the meetings and related research provided by the national fisheries management department or agency. Rarely are people found from environmental departments on its fishery commission.

Adding to that, the interests of individual members, including the social and economic needs of industry groups, often dominate the decisions and delay actual decision-making, as noted by the performance review. A good example of this is seen in fisheries' management on the eastern Atlantic and the Mediterranean Bluefin tuna.

According to the CCSBT performance review (2008), the current process for developing and providing scientific advice on SBT from the Extended Scientific Committee to the CCSBT is an excellent model that has helped improve the integrity of the CCSBTs scientific process. Access to highly competent national scientists has been available and is reflected in the abundance and quality of scientific papers presented to the various CCSBT scientific forums. The independent panel and chair arrangements have added, in 1999, further support to this process and militated against the tendency of member scientists to modify their advice for reasons associated with their national interests.

However, the CCSBT has specific provisions within its procedure rules for making documents presented to meetings confidential. Yet public access to scientific documents used in public decision-making is considered a fundamental principle under which science is conducted within RFMOs (e.g., UNFSA) and more generally within many societies (e.g., freedom of information laws) (Polacheck 2012).

There is no doubt that the lack of transparency, in this case, is misrepresentative of the information and destroys a unique chance of creating a knowledgeable and trusted science-based policy.

What is notable in looking into the scientific bodies of CCAMLR, ICCAT, and CCSBT are the very high caliber scientists; the difference between them is the level of a country's information dependence. The system's robustness might be increased by reducing this dependency through greater collective investment in scientific programs to collect and compile independent information.

Conclusion

Considerable attention has been devoted to studying how RFMOs face up to such challenges to become even more effective (Molenaar 2003, 2007). As highlighted by Polacheck (2012), the independence and separation of the scientific processes that provide information for consideration by the RFMO from the policy deliberations of the RFMO is an important component influencing the intent of UNCLOS and UNFSA. Their management decisions are based on the best available scientific information.

There has been an increasing recognition in recent years of the need for RFMOs to improve their performance because of demands contained within current international agreements aimed at better conservation and management of fishery resources. To do that, an insulated scientific body is essential. The choice, quality, and diversity of the expertise involved in working groups and the scientific community depend on their members' contributions and engagement to a considerable extent.

The RFMOs institutional design analyzed in this chapter does not fully allow the scientists to produce usable knowledge, as defined by Clark and Majone (1985) and demonstrated empirically in the Haas and Stevens (2011) chapter. The science is not fully safeguarded enough from politics, which may, despite the quality of science produced, not be considered accurate by many decision-makers.

However, the next chapters will evaluate specific cases where the situation can be overcome as it depends on an array of diverse factors. In addition, even still needing improvement, CCAMLR presents the most established collection of factors to ensure that knowledge may contribute to power and that, along with a higher level of social learning, maybe one of the factors guaranteeing a higher biomass recovery performance, as shown by Cullis-Suzuki and Pauly (2010). Conversely, ICCAT and CCSBT need to find a path towards protecting their scientific bodies to produce more independent, accurate, and legitimate information to improve their social learning and, consequently, the effectiveness in recovering fish populations.

This chapter agrees with Haas and Stevens (2011) that knowledge can speak volumes to power, and with this comes more effectiveness in terms of problem-solving. However, to make it work for fishery agreements, expertise and claims are developed behind a politically insulated wall, a wall that is not yet working properly in every situation or every time the decision is made.

Epistemic communities for these agreements need to be fully understood, and knowledge must also possess the substantive characteristics of usable knowledge: credibility, legitimacy, and saliency, to influence the RFMO's effectiveness.

In the following chapters, an effort will be undertaken to investigate further the CCAMLR, ICCAT, and CCSBT epistemic communities, including who they are, how they act, and how they play a role in these specific RFMOs, as yet another component towards enhancing their effectiveness.

Chapter 2
CCAMLR: To Be or Not To Be an RFMO?

Introduction

The Southern Ocean surrounds Antarctica and represents approximately 15% of the world's ocean area. The Antarctic region constitutes a fragile ecosystem closely related to the unique features of that continent's physical environment. Despite its forbidding climate and isolation from the rest of the world, Antarctica—which once fascinated the early explorers—has always been the center of unprecedented international scientific research and cooperation. The Antarctic Treaty (Article IX) established that the Southern Ocean must be a place devoted "to peace and science."[1]

As described in detail by Shusterich (1984), since the frozen continent was discovered, several claims on the sovereignty of its territory were debated. And several attempts to create international control over Antarctica were made in the 1950s. For instance, in 1956, the permanent representative from India to the United Nations requested that "the question of Antarctica" was to be included on the provisional agenda of the United Nations General Assembly. India's goals were to ensure that an international agreement could consider the development of Antarctic resources only for peaceful purposes, that the area would be non-militarized, that nuclear weapons testing would be banned, and that future disputes would be referred to the International Court.

The proposal was withdrawn mostly because of opposition from Chile and Argentina and lack of support from the United States and the Soviet Union—countries with high interests in the sovereignty of the Antarctic area. The idea of a United Nations trusteeship for Antarctica was introduced the same year (and again in 1958) by Prime Minister Nash of New Zealand. He stated that any arrangement for

---

[1] Due to the ATS, many NGOs use the expression to protect the Southern Ocean: http://www.greenpeace.org/international/en/campaigns/oceans/our-oceans-and-seas/southern-ocean/; http://www.pewtrusts.org/en/places/southern-ocean; https://www.bas.ac.uk/about/antarctica/the-antarctic-treaty/the-antarctic-treaty-explained/.

© The Author(s), under exclusive license to Springer Nature Switzerland AG 2021
L. R. Gonçalves, *Regional Fisheries Management Organizations*,
https://doi.org/10.1007/978-3-030-70362-2_2

international control of Antarctica should have the approval of the United Nations (Shusterich 1984).

Even with all of those sovereignty interests, science played a key role in the previous events that led to the Antarctic Treaty. During the International Geophysical Year (IGY) of 1957/1958, 12 nations (the seven claimants and the United States, Soviet Union, Belgium, Japan, and South Africa) implemented major Antarctic scientific research programs (Shusterich 1984).

At the end of the IGY, there was general agreement among the 12 nations that had participated in the science projects that the international cooperation level should be continued. The United States took the initiative on 2 May 1958 when it proposed to the other IGY participants that they should all join in a treaty designed to preserve the continent as a space safeguard to scientific research and for peaceful purposes (Berkman et al. 2011).

The Conference on Antarctica was held from 15 October to December 1959, and the outcome of the Washington negotiations was the so-called Antarctic Treaty. The negotiations proved successful amid the Cold War between the United States and the Soviet Union. They were an early indication that both superpowers were prepared to consider Antarctica a special case in international politics (Shusterich 1984).

The Treaty was signed in 1959 and entered into force in 1961 following ratification by all 12 original signatories. The original signatories—Argentina, Australia, Belgium, Chile, France, Japan, New Zealand, Norway, South Africa, the United Kingdom, the United States, and the Soviet Union—formed the initial membership of the management group known as the consultative parties.

Since 1959, 40 new countries have entered the Treaty. These countries may participate in the Consultative Meetings. Seventeen of the acceding countries have had their activities in Antarctica recognized, and consequently, there are now 29 Consultative Parties in all: Czech Republic (1962), Brazil (1983), Bulgaria (1978), China (1983), Ecuador (1987), Finland (1984), Germany (1979), India (1983), Italy (1981), Korea (1986), Netherlands (1967), Peru (1981), Spain (1982), Sweden (1984), Ukraine (1992), and Uruguay (1980). The other 23 non-consultative parties are invited to attend the consultative meetings. Still, they do not participate in the decision-making.[2]

Thus, in this context, in the middle of wartime, the Antarctic Treaty was signed to guarantee peace and science for the frozen continent. The Antarctic Treaty has demonstrated considerable adaptability and resiliency as it evolved from a single instrument into a robust regional regime containing four new instruments since its inception, known since then as the Antarctic Treaty System (ATS, hereafter): the 1964 Agreement includes measures for the protection of flora and fauna, the 1972 Convention on the Conservation of Antarctic Seals, the 1980 Convention on the Conservation of Antarctic Marine Living Resources (CCAMLR), and the 1991

---

[2] From ATS website membership update list—http://www.ats.aq/devAS/ats_parties.aspx?lang=e (Accessed in June 2015).

Protocol on Environmental Protection to the Antarctic Treaty (Environmental Protocol) (Joyner 2011).

This chapter will be focused on CCAMLR's development and its functioning related to the role of epistemic communities; however, the history of ATS could not be set aside, as CCAMLR was created under the ATS, which itself possesses different characteristics from other RFMOs. CCAMLR will be evaluated as part of ATS and as an independent agreement, with its provisions and norms. And the main focus will be how science and the epistemic communities are influencing policies.

CCAMLR: An Agreement for Conservation of Marine Living Resources

The Antarctic Treaty stands alone as an instrument of conflict prevention, strategic accommodation, and political cooperation, mostly because of sovereignty. Most notably, Articles II and III of the Antarctic Treaty provide "freedom of scientific investigations" and promote "international cooperation in the scientific investigation." Both are considered essential elements for promoting peace, cooperation, and the progress of all humankind (Miller 2011).

Although the Treaty contains specific provisions to preserve and conserve the Antarctic living resources (article IX, para 1f), it was general enough not to regulate or impose any fishery restriction. Thus, the extensive fishery of finfish[3] and large-scale krill exploitation still raise concerns about fisheries sustainability in the Treaty area (Miller 2011).

In response, Recommendation VIII-10 from the 1975 Eighth Antarctic Treaty Consultative Meeting noted the need to "promote and achieve within the framework of AT, the objectives of protection scientific study and rational use of Antarctic marine living resources and, then, the importance of science was recognized as a basis for the protection and rational use of such resources" (Berkman et al. 2011).

According to Berkman et al. (2011), in 1977, the Scientific Committee on Antarctic Research (SCAR) and the Scientific Committee on Oceanic Research (SCOR) sponsored the Biological Investigations on Marine Antarctic Systems and Stocks (BIOMASS) intending to "gain a deeper understanding of the structure and dynamic functioning of the Antarctic marine ecosystem as a basis for future management of potential living resources" (El-Sayed 1994). BIOMASS and three more reports produced under the United Nations Food and Agriculture Organization (Eddie 1977; Everson 1977; Grantham 1977) highlighted that krill is a keystone species Antarctic marine ecosystem.

---

[3] Finfish is a real fish, such as bony fish or a cartilaginous fish, especially in contrast to a shellfish or other aquatic animal.

As indicated by Miller (2011), growing recognition of krill's ecosystem role increased the concerns that its large-scale exploitation could have severe repercussions for Antarctic birds, seals, and whales.

At the same time, Recommendation IX-2 from the Ninth Antarctic Treaty Meeting (1977) called the Parties to contribute to scientific research on Antarctic marine living resources, observe interim guidelines on their conservation, and schedule a special meeting to establish a conservation regime for these resources.

The Second Special Antarctic Consultative Meeting comprised a series of meetings from 1978 to 1980 and concluded with the signing of the Convention on the Conservation of Antarctic Marine Living Resources (CCAMLR 1980). The Convention entered into force on 7 April 1982.

Although developed under the Antarctic Treaty, the CCAMLR stands alone as a legally binding agreement, and its attached Commission has its own personality (Miller 2011).

The CAMLR Convention Area encompasses approximately 10% of the global ocean area. However, the extended area of application of the CAMLR Convention included the territories of various sub-Antarctic Island groups and their maritime jurisdictions (Fig. 2.1).

The adoption of CCAMLR was a significant step-change in the development of the ATS. The Convention's primary objective was the conservation of marine living

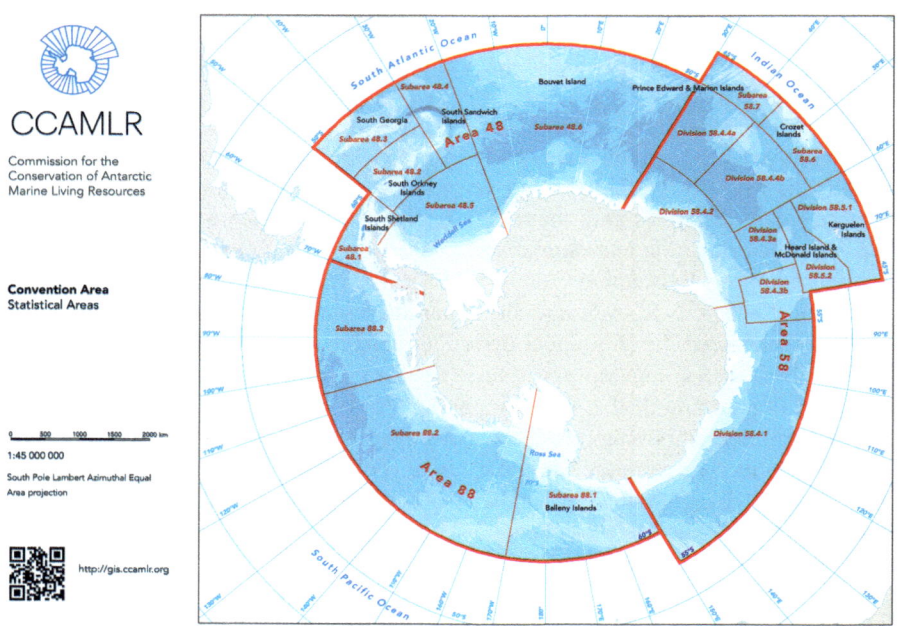

Fig. 2.1 CCAMLR Convention Area (Source: CCAMLR (2017)

resources, but with the understanding that conservation includes rational use. This emphasis on conservation, but with harvesting seen as integral to that principle, remains a fundamental provision of CCAMLR and one that continues to set it apart from the more traditional RFMOs with their focus on the management of target fish stocks. Yet, as history shows, CCAMLR was built based on the involvement of science, portraiting as a leader for the sustaianble management of marine living resources (Miller 2011; Willock and Lack 2006).

Why CCAMLR Is Widely Recognized as an Example to Other RFMOs

Some stakeholders that are following up or working on CCAMLR hardly consider it an RFMO. Rather, CCAMLR is widely recognized as a progressive international commission responsible for conserving the Southern Ocean marine ecosystem while having within it the attributes of an RMFO (Constable et al. 2000; Bodin and Österblom 2013).

In its Convention and established practice, it is well known as a leading organization in developing best practices in the ecosystem approach to managing activities in waters beyond national jurisdiction (Mooney-Seus and Rosenberg 2007; Ruckelshaus et al. 2008). CCAMLR has notably achieved these advances without precedent, only guided by the principles within Article II of its Convention. It is likely its history and subsequent development that made this organization behave differently from other RFMOs.

Article 1(1) of CCAMLR states that the Treaty's objective is the conservation of Antarctic marine living resources; however, "conservation" is defined to include "rational use." Accordingly, CCAMLR addresses both species protection and their "rational use." CCAMLR, therefore, has potentially conflicting goals of exploitation and conservation of marine species in Antarctica. In this respect, through its institutional bodies (i.e., the CCAMLR Commission, the Secretariat, and Scientific Committee), CCAMLR acts as a regional fisheries organization that manages living resources in the Southern Ocean area (Bender 2007).

The negotiators of CCAMLR are constituted by the Antarctic Treaty Contracting Parties (ATCPs, hereafter) and the inaugural members of CCAMLR. The original membership of eight states was extended by adding the other seven original signatory states that subsequently ratified the Convention and became members of the Commission. Since then, other countries have acceded to the Convention, with many states also gaining membership of the Commission. To date, there are now a total of 25 members of the Commission. Eleven other states have acceded to the Convention but are not members of the Commission (Table 2.1).

In effect, the members have responsibility for the executive functions of CCAMLR (including the adoption of Conservation measures) and contribute to the organization's budget (principally for the running of the Secretariat). The acceding

Table 2.1 Members and acceding states of CCAMLR to date (Source: CCAMLR website, accessed in June 2015)

	Members	Acceding states
01	Argentina	Bulgaria
02	Australia	Canada
03	Belgium	Cook Islands
04	Brazil	Finland
05	Chile	Greece
06	China	Mauritius
07	European Union	Netherlands
08	France	Islamic Republic of Pakistan
09	Germany	Republic of Panama
10	India	Peru
11	Italy	Vanuatu
12	Japan	
13	Republic of Korea	
14	Namibia	
15	New Zealand	
16	Norway	
17	Poland	
18	Russian Federation	
19	South Africa	
20	Spain	
22	Sweden	
23	Ukraine	
24	United Kingdom	
25	Uruguay	

states are known as Contracting Parties (CPs), which are not Members of the Commission, but they are signatories to the Convention. They do not pay dues and do not even participate in CCAMLR meetings, and in contrast, are not party to decision-making, nor are they liable for subscription costs. Such states are invited as observers to the annual meetings of CCAMLR. All CPs (both members and the acceding states) are nevertheless bound by the obligations of relevant Conservation Measures (described further below).

Like most other fishery agreements, the CCAMLR sets forth a general purpose and establishes a Commission to effectuate that purpose. In formulating its conservation measures, the Commission is assisted by a Scientific Committee charged with collecting and analyzing data and suggesting to the Commission its recommended conservation methods (managing advice). The relevance of its Scientific Committee is clearly expressed on the CCAMLR structure, wherein the Scientific Committee is providing not only advice to the Commission, but is contained within the Commission. As stated on their website, "The Commission includes a Scientific Committee established by the CAMLR Convention" (Fig. 2.2).

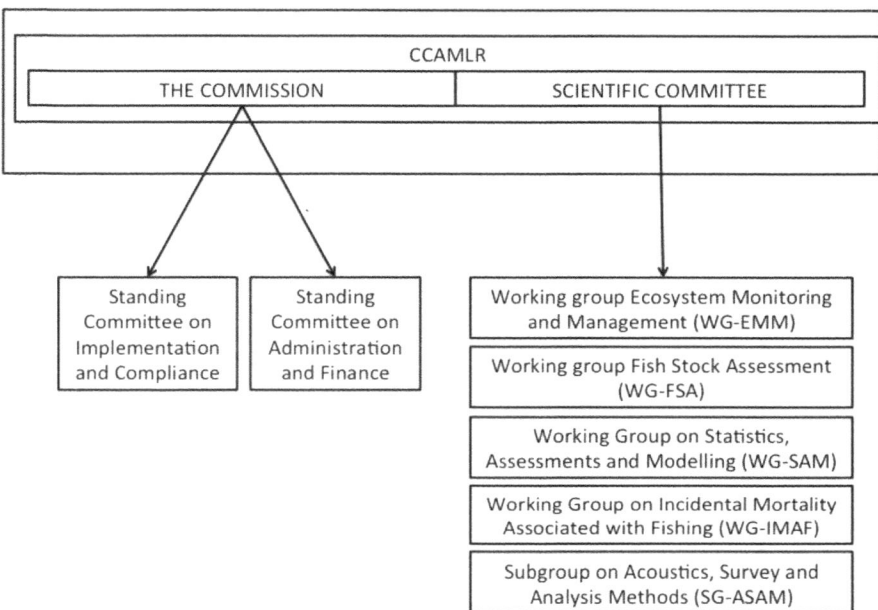

Fig. 2.2 CCAMLR structure

CCAMLR has often been referred to as "something more than an RFMO" (CCAMLR 2008). Underpinning this statement has been CCAMLR's integral position within the ATS and strong legal linkages to the Antarctic Treaty. Furthermore, the overarching objective of the Convention has been the conservation of marine living resources. These two aspects have set CCAMLR apart from the more traditional RFMOs with their emphasis on the harvesting of commercial target species.

The strong conservation credentials of CCAMLR, along with the precautionary principles and ecosystem approach embodied within the Convention, have enabled CCAMLR, at times, to take the lead in developing management tools with a strong emphasis on conservation and sustainability—the trade-related Catch Documentation Scheme (CDS), and the mitigating measures for seabirds are but two examples of where CCAMLR has developed best practices in international fisheries management terms.

According to the CCAMLR (2008), the distinction between CCAMLR and RFMOs has lessened in recent years. The reasons for this are varied but include:

(a) the changing emphasis within CCAMLR of the ratio of fishing to non-fishing Members of the Commission.
(b) the increasing numbers of CPs that have no traditional linkage with the ATS.
(c) the increasing trend for CCAMLR members to be represented at Commission meetings by officials from Fisheries Ministries rather than from Ministries of Foreign Affairs (where responsibility for the ATS usually resides).

Table 2.2 Percentage of countries with interest in fishing at CCAMLR area

Year	Total member	No of fishing members	% Fishing states
1985	16	6	38
1995	22	9	41
2005	24	16	67
2015	25	17	68

(d) that the ecosystem approach and the precautionary principle have also been adopted by some RFMOs, which is a good point.

At the time of entry into the force less than 40% of the members were fishing states. That proportion has increased over time with, by 2005, almost 70% of members fishing (see Table 2.2 below updated from CCAMLR 2008).

One characteristic essential to consider is the RFMO's composition—the parties, members, observers, and lobbyists who convene around the issues subject to the RFMO. If an organization has a biased set of participants (e.g., being comprised only of users), the outcome is likely to reflect that bias (e.g., increased harvest levels). As stated before, one reason for CCAMLR's success in passing more ecosystem-based and conservation measures is that it is comprised of both fishing and non-fishing science-focused members. Therefore, one institutional characteristic that may prove essential in effectively and sustainably managing marine living resources is that of a stakeholder base that extends beyond those exploiting the resource. In this regard, transparency and access to non-fisheries-dependent data are essential as lack of adequate access to information can limit the ability to peer-review the scientific advice, participate in rulemaking, and track compliance (Brooks et al. 2014; Ardron et al. 2013).

Additionally, it is important to highlight that two provisions of CCAMLR, which were highly innovative at the time of its adoption, remain key to its current approach and have been adopted more recently by RFMOs—namely the precautionary principle and the ecosystem approach. Those approaches made a significant difference in the management system.

With only these two points, it is already possible to detect differences in CCAMLR's history from many other RFMOs such as ICCAT and CCSBT. Now, RFMOs adopt those same measures; CCAMLR was created in this spirit—learning from doing. The agreement emerged in a different context by considering the need for scientific cooperation and conservation guidelines to deal with depleted stocks.

For most RFMOs, the traditional approach to management has been the maximum sustainable yield concept, aiming for that level of harvesting, which will maximize the catch of the species plotted over a time series of estimated species productivity. In the Southern Ocean, however, the applicability of this oft-used concept has been seriously challenged because of the fishery's peculiar attributes and the proposed ecosystem approach of the Antarctic Living Resources Convention. The United States delegation spearheaded the fight for adopting a "multispecies approach" towards managing the Southern Ocean during the negotiations (Frank 1983).

The Convention gives the Commission the power to formulate, adopt, and revise conservation measures pertaining to particular species. In this respect, the Commission maintains an "ecosystem approach" to fisheries management. This approach envisions managing marine living resources by examining the effect of maintaining a particular population and harvesting levels on the entire ecosystem. The ecosystem approach also uses "feedback management." Using this technique, scientists set a species-specific target population and monitor changes from that target. "If the actual population level begins to deviate from this target, various management control techniques of the system can be altered to maintain the target population."

Science and the Antarctic

International interest of a scientific nature in the Antarctic Region commenced with the era of exploration, continues through today, and accounts for some of the basis of the region's legal regime. Every State that laid claims to Antarctic territorial sovereignty and every interested non-claimant state conducted scientific operations between the World Wars. It is likely that many scientific expeditions to Antarctica had secondary motivations—the perfecting of claims to territorial sovereignty. Sovereignty claims, in turn, form the basis for the perceived State's rights over Antarctic natural resources. The United States and the Soviet Union conduct marine research all around the continent while other interested states engage in more modest operations. Although much of this research was oceanographic or meteorological, a substantial portion of recent scientific efforts has been devoted to evaluating the Southern Ocean's prospects as an area for fisheries (Frank 1983).

The management on CCAMLR appears to comply consistently with scientific advice and corresponding management measures (Mooney-Seus and Rosenberg 2007); this is proved by Table 2.3. When the Scientific Committee reaches consensus in their meetings, the advice is hardly discussed on the plenary as it is usually accepted as a matter of fact. As we can learn from Chap. 1, the entire institutional

Table 2.3 Catches, TAC recommended, Management advice, and Estimate of IUU in CCAMLR South Georgia and Shag Rocks (Subarea 48.3) for Patagonian toothfish (*Dissostichus eleginoides*)[a]

D. Eleginoides	2004	2005	2006	2007	2008	2009	2010	2011	2012	2013	2014
Catches	4497	3034	3535	3539	3864	3382	2519	1763	1806	2094	2180
TAC rec	4420	3050	3556	3554	3920	3920	3000	3000	2600	2600	2400
Management advice	b	3556	3554	3920	3920	3000	3000	3000	2600	2600	2400
Estimate of IUU	0	23[c]	0	0	0	0	0	0	0	0	0

[a]The Management advice provided by the Scientific Committee entry in force in the year to follow
[b]Scientific Committee was unable to provide a number
[c]An additional 23 tons was taken by a single IUU vessel (the Elqui) apprehended by the UK prior to the fishery

design of CCAMLR differs from the other RFMOs included in this research, which efficiently contributes to the flow of knowledge from science to the Commission itself.

What accounts for that is that the CPs are following the rule that was established by Article IX of the CAMLR Convention text, which says:

> The function of the Commission shall be to give effect to the objective and principles set out in Article II of this Convention. To this end, it shall:
>   (f) formulate, adopt and revise conservation measures on the basis of the best scientific evidence available, subject to the provisions of paragraph 5 of this Article;

And in Paragraph 2 highlights the scope of the Conservation Measures (CMs):

> 2. The conservation measures referred to in paragraph 1(f) above include the following:

(a) the designation of the quantity of any species which may be harvested in the area to which this Convention applies;

(b) the designation of regions and sub-regions based on the distribution of populations of Antarctic marine living resources;

(c) the designation of the quantity which may be harvested from the populations of regions and sub-regions;

(d) the designation of protected species;

(e) the designation of the size, age, and, as appropriate, sex of species which may be harvested;

(f) the designation of open and closed seasons for harvesting;

(g) the designation of the opening and closing of areas, regions, or sub-regions for purposes of scientific study or conservation, including special areas for protection and scientific study;

(h) regulation of the effort employed and methods of harvesting, including fishing gear, with a view, inter alia, to avoiding undue concentration of harvesting in any region or sub-region;

(i) the taking of such other conservation measures as the Commission considers necessary for the fulfillment of the objective of this Convention, including measures concerning the effects of harvesting and associated activities on components of the marine ecosystem other than the harvested populations.

As part of its approach to fisheries management, CCAMLR has released several conservation measures aimed at the conservation and rational use of certain species and fisheries practices in general. These conservation measures include "precautionary" fisheries catch limits for particular species. Such limits are conservative catch limits considering the scientific uncertainty surrounding specific species, including population levels, recruitment rates, and interactions with other species (Bender 2007).

To illustrate this management process, this research evaluated the development and input made by the Scientific Committee on toothfish management from 2004 to 2014. It analyzed how the Commission took the decisions. The compiled data were mainly for Subarea 48.3 (Fig. 2.1).

The Patagonian Toothfish

The Patagonian toothfish (*Dissostichus eleginoides*) and the Antarctic toothfish (*Dissostichus mawsoni*) are two distinct Antarctic cod families species. Both species are legally commercially fished in the Southern Ocean and are sold on the international market, most commonly, like Chilean sea bass. The CCAMLR is the RFMO responsible for managing fisheries in the Antarctic and, until 1998, no distinction was made between the two species for management purposes. Both species were accounted for under statistics compiled for Patagonian toothfish. The commercial fishery in the Southern Ocean has primarily targeted the Patagonian toothfish.

This research will look into the *D. eleginoides* data to see the influence of science and possibly the emergence of an epistemic community to establish rules of fishery management.

The Patagonian toothfish (*D. eleginoides*) is a large, long-lived species belonging to the family *Notothenidae* or Antarctic cods. Toothfish show distinct depth preferences with age, with juveniles (<50 cm) living on the continental shelf and moving into deeper water (>500 m) as they reach maturity (~90 cm). Toothfish are essential predators, feeding primarily on fish, cephalopods, and crustaceans; they also scavenge (CCAMLR 2014; fishery report).

In the early 1990s, the collapse of fish stocks in many other fisheries worldwide displaced many fishing fleets. As these fleets sought other valuable fish stocks, the large size and delectable meat of the Patagonian toothfish quickly caught the attention of the international market (CCAMLR 2014).

By the mid-1990s, Patagonian toothfish was dubbed "white gold," and it is now considered one of the most valuable fish species on the market. Prices have been as high as US$10 per kilo for headed, gutted, and tailed fish in the main markets in the United States, Japan, Europe, Canada, and, increasingly, Asia and China (Isofish 2002; TRAFFIC and WWF 2002; Lack 2001; Kock 2001; Dodds 2000; ASOC 1998; Perry 1998). Over 90% of toothfish products are sold internationally, mainly to Japan, the United States, Europe, and China (TRAFFIC and WWF 2002).

This high valuation led to a seemingly overnight intensification of the fishery and placed more significant pressure on the fish stocks. CCAMLR grew increasingly concerned, as toothfish have a long life span, late sexual maturity, and low fertility; thus, they are particularly vulnerable to overexploitation. Despite CCAMLR's attempts to implement management measures, the high market value of the fish and remoteness of the fishing grounds resulted in substantial (Riddle 2006) overfishing.

Catches of *D. eleginoides* in Subarea 48.3 were initially reported in 1977. Until the mid-1980s, the fishery was carried out entirely by bottom trawls.[4] The longline fishery probably began in April 1986 (WG-FSA-92/13). The annual catch data are summarized in Table 2.3.

---

[4]A bottom trawl is constructed like a cone-shaped net towed/dragged (by one or two boats) on the bottom. It consists of a body ending in a codend, which retains the catch.

To manage the Patagonian toothfish fishery, CCAMLR took over a series of management measures. First, they established a regulatory framework that considers different types of fisheries.

A "new fishery" exists when biological data (including species distribution and abundance) and fishery data are not available yet, or data from the two most recent fishing seasons have not been submitted to CCAMLR. In this scenario, notification is required prior to fishing (Conservation Measure 21-01[5]) and it becomes an exploratory fishery after the first year of fishing.

The "exploratory fisheries" are not allowed to expand faster than the acquisition of information necessary for managing the fishery within CCAMLR's management objectives. For this type, notification and permission are required before fishing (Conservation Measure 21-02[6]) and it remains an exploratory fishery until sufficient information is available on appropriate catch and effort levels and the potential impacts on dependent and related species.

The "established fisheries" category is related to fisheries that have been in progress for several years and for which assessments are available that are sufficient to directly estimate stock size, stock status, and the catches consistent with achieving management objectives. Some of the toothfish fisheries are treated this way, and for these fisheries, stock conditions and assessments are typically reviewed and revised annually or bi-annually. Assessments take account of all sources of fishing mortality, including estimated IUU catches. Notification and permission are required prior to fishing for krill (Conservation Measure 21-03[7]), and are optional for other target species.

Finally, the last two are "lapsed fishery," when fishing operations have ceased due to commercial considerations, and assessments are no longer current, and "closed fishery," when directed fishing on the target species is prohibited.

With this, they guarantee appropriate and specific management to different fish populations and stocks.

The assessments were, and at a certain level, still are, under a high level of uncertainty. Because the Antarctic is a remote area, there is still a lack of knowledge about its ecosystem. In addition, because the approach used by CCAMLR is precautionary and ecosystemic, which has some advantages, it is still new for scientific methods. This is one reason to justify why the Scientific Committee does not always reach consensus in their decisions, as mentioned previously in Chap. 1.

So, in cases where the Scientific Committee is not confident, the Commission must decide on its own. For instance, in 1994, the Commission noted that the assessment methods, which had previously been judged satisfactorily, had been invalidated by WG-FSA at its meeting. Thus, the Scientific Committee had been unable to recommend an appropriate TAC level for this fishery (SC-CAMLR-XIII, paragraph 2.29).

---

[5] https://www.ccamlr.org/en/measure-21-01-2010.

[6] https://www.ccamlr.org/en/measure-21-02-2013.

[7] https://www.ccamlr.org/en/measure-21-03-2014.

In this specific case, the Commission considered that its ability to formulate conservation measures based on objective scientific analysis and advice was fundamental to its work and the Convention. In this regard, the Commission strongly endorsed the intention of the Scientific Committee to hold a workshop just prior to the meeting of WG-FSA in 1995 on the development of methods for assessing the biomass of *D. eleginoides* (SC-CAMLR-XIII, paragraph 2.171995).

The Workshop on Methods for the Assessment of *D. eleginoides* (WS-MAD) was held at CCAMLR Headquarters, Hobart, Australia, from fifth to ninth October 1995. The Workshop's main aim was to develop methods for assessing the biomass and status of *D. eleginoides* stocks. The full terms of reference for the Workshop are given in SC-CAMLR-XIII, paragraph 2.17.

Circumstantial evidence and confidential records indicate that the reported catches of *D. eleginoides* by longliners in Subarea 48.3 and adjacent banks do not represent the accurate removal level. Since many of the methods of estimating the abundance of *D. eleginoides* rely on estimates of total removals, the Workshop agreed that every effort should be made to estimate these as accurately as possible.

At its 1995 meeting, the Working Group had noted that the reported catch for *D. eleginoides* probably represented only about 40% of the total removals from the fishery. Since the total removals is an essential component of any assessment, this level of uncertainty had been viewed with considerable concern (CCAMLR 1996, p. 7).

With the passing years, the Commission and its Scientific Committee adjusted its methods and information to provide the best scientific advice they could, with full support by its CPs.

It is crucial for stock assessment to have as complete information as possible on removals of fish from a stock. A large number of Commission circulars (COMM CIRCs 96/71, 97/4, 97/26, 97/27, 97/38, 97/40, 97/43, 97/48, and 97/50) drew attention to high levels of unregulated fishing on *D. eleginoides*.

In a 1997 meeting, the Chairman opened the meeting by stating that "these progressive measures had little effect if they were not effectively implemented. The extent of illegal fishing had led to great concern and had visibly undermined the conservation policy of CCAMLR" (1997, p Opening). The stocks of *D. eleginoides*, in particular, were under pressure because of illegal fishing. The issue of illegal fishing—and measures to contain it—was a serious issue facing the Commission at this meeting, and central to this containment were measures of control and enforcement.

The Chairman further commented that the extent of illegal fishing, particularly for Patagonian toothfish (*D. eleginoides*), has led to great concern and seriously undermines CCAMLR's conservation policies.

The main countries legally harvesting and exporting Patagonian toothfish were Argentina, Australia, Chile, France, South Africa, and the United Kingdom, and they supplied over 70% of the legal market in 2000 (TRAFFIC 2001). Countries that are or have been involved in IUU fishing or trade, either as flag states, countries of vessel ownership, nationality of the master, or ports of landing, have been identified as including Russia, Argentina, Chile, Spain, Uruguay, Belize, Denmark, Mauritius, Namibia, Panama, Sao Tome, and Principe, Seychelles, Vanuatu, St.

Vincent, and the Grenadines, Tongo, Indonesia, and China (Austral Fisheries 2002; Lack and Sant 2001). Also, although Argentina, South Africa, and the United Kingdom legally harvest toothfish, these countries have also been identified as having been involved in IUU fishing activities.

Major importers of toothfish include Japan and the United States, although Canada and the European Union also import toothfish and, increasingly, China is a market. Those countries identified as being involved in IUU fishing or trade are parties to CITES, and Argentina, Chile, Japan, Namibia, Norway, Russia, South Africa, Spain, the United Kingdom, and Uruguay are CCAMLR members. However, countries involved in IUU fishing are constantly changing, and when some countries register vessels, they may be unaware that their ships are engaged in IUU fishing of Patagonian toothfish.

IUU fishing for Patagonian toothfish has compromised the effectiveness of CCAMLR conservation measures and is allegedly at a scale that threatens the sustainability of regulated fisheries and the survival of albatross species (ISOFISH 2002; ASOC 2002; Greenpeace 2000).

Aiming to address illegal fishing, the Catch Development Scheme (CDS) was adopted in 2000. The CDS is designed to track Patagonian toothfish' landings and trade flows in its area and restricts access to markets for toothfish caught by IUU fishing. The scheme enables the Commission to identify the origin of toothfish entering all parties' markets to the scheme and helps determine whether the fish are caught in a manner consistent with CCAMLR provisions. The system requires specific control by port states. A fishing vessel must provide prior notification of its intention to enter a port, including a declaration that it has not been involved in IUU fishing. The vessel's flag state shall also confirm this declaration, and those vessels failing to make a declaration shall be denied port access. If there is evidence that the ship has fished in contravention of CCAMLR conservation measures, the catch shall not be allowed to be landed or transshipped (Lodge et al. 2007).

This IUU threat has been effectively addressed with the new norms and rules. Except for just a few years and punctuated occasions, the Commission has steadily decreased illegal, unregulated and unreported fishing over the past twenty years. As a consequence, illegal, unregulated, and unreported fishing has been reduced to less than 10% of its peak value in the mid-1990s (Österblom and Sumaila 2011; Österblom and Bodin 2012). During the years of this research, there is no evidence of illegal, unreported, and unregulated (IUU) fishing since 2006.

The CCAMLR Epistemic Community

According to Lodge et al. (2007), there is a "lack of political will by fishery managers and marine resource users to implement management measures according to scientific advice and effectively enforce and comply with those management measures." However, this does not seem a problem for CCAMLR, as it is shown in Table 2.3.

During 10 years of evaluation for toothfish populations on zone 43, one of the most productive zones for toothfish fishery, all management advice proposed by the Scientific Committee was accepted by the Commission, including designing the CDS.

According to this research, at CCAMLR, an epistemic community could be identified under the Working Group Fish Stock Assessment (WG-FSA) (see Fig. 2.2). The group is formed mostly by high caliber scientists operating under their knowledge area and are in a condition to offer the best scientific advice to the Commission.

It is under the WG-FSA that decisions about management advice related to fish stock assessment are made. Throughout the 10 years evaluated, all decisions accepted by the Commission became conservation measures.

The Working Group noted in the WG-FSA report for 2015 (p. 333) that, while the median SSB[8] (Fig. 2.3) was estimated to have fallen below the target level of 50% of the pre-exploitation median SSB from 2009 to 2012, it was above the target level in 2015. It did not fall below the target for the remainder of the projection period under the recommended yield (paragraph 4.37, should be set at 2750 tons, more than 2400 from the last year).

This type of information from the WG-FSA is used to subsidize the decisions of the Scientific Committee. The Scientific Committee meets annually immediately prior to the Commission meeting. However, information is produced in the working groups, and there was no influence from the Scientific Community on the results of the WG-FSA.

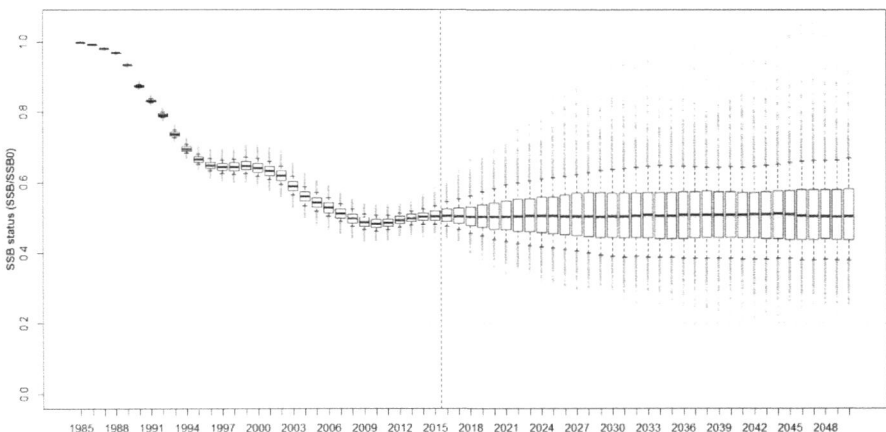

Fig. 2.3 *Dissostichus eleginoides* in Subarea 48.3 SSB status estimated by the model described in WG-FSA-15/59. Dashed horizontal lines show a status of 0.5 and 0.2. Source WG-FSA report from 2015, p. 378

---

[8] SSB—Spawning stock biomass. Total weight of all sexually mature fish in the stock.

During the 10 years of this research, 144 representatives were part of the WG-FSA. However, only 25 of them would be identified as members of the influential epistemic community for the decisions around the management advice for the toothfish. They were identified through the snowball technique and by evaluating the participants' list and their peer-review publications.

As members of this community, they could produce usable knowledge (Haas and Stevens 2011) accurate enough to influence the Commission's decision.

Unlike ICCAT, where the institutional design for science is not well established in terms of procedures (as explained in the previous chapter), in the CCAMLR case, the epistemic community can act and influence the policy decisions mediated by the institutional design. They do not need to persuade the CPs, nor is it necessary to arrange critical players to make the knowledge acceptable. They simply work as scientists; with reliable information, they have at hand and produce usable knowledge good enough to be accepted. The only matter here is when scientists in the Working Group cannot reach a consensus for management advice. In these specific situations, the Commission makes the decision, as usable and consensual knowledge could not be produced.

The members of the Working Group and the Scientific Committee possess the expertise necessary to understand the issues at stake, to interpret the information similarly, and then to form the same goals and shared beliefs about what should be done in terms of policy based on the scientific information they have. Under the WG-FSA, new methodologies emerge to overcome the lack of certainty on the information (Cross 2015).

The members ($N = 25$) here indicated (Annex I) have all the characteristics indicated by Haas (2014) to identify the members of an epistemic community. They share knowledge about the causation of social or physical phenomena in an area for which they have a reputation for competence and a common set of normative beliefs about what actions will benefit human welfare in such a domain.

NGOs at CCAMLR

The involvement of NGOs in CCAMLR is also different from other RFMOs.

In addition to governments and the CCAMLR Commission's responsibility, some non-governmental players monitor Patagonian toothfish management. For example, the Antarctic and Southern Ocean Coalition (ASOC) is an NGO established in 1976/1977 to coordinate activities concerning Antarctica and its surrounding oceans and to provide input into the Antarctic Treaty System (ATS), including CCAMLR (Boyd 2002; Wapner 2000). ASOC comprises over 250 conservation groups from more than 50 countries. Its strength stems from this alliance, its relative singularity of purpose, and its ability to draw upon many contacts and access to governments.

The Antarctica Project is the Secretariat of ASOC and is based in Washington, D.C. Before entering the IUU fishing debate, ASOC campaigned to "protect the biological diversity and pristine wilderness of Antarctica, including its oceans and marine life to ensure that the environment comes first" when decisions were made under the Antarctic Treaty (ASOC 2001). During the 1980s, ASOC played a role in the discussion that led to the Convention on the Regulation of Antarctic Mineral Resource Activities (CRAMRA) and its eventual replacement by the 1991 Protocol on Environmental Protection to the Antarctic Treaty. Since then, ASOC has contributed to the ATS's ongoing development and its associated environmental protection measures.

In 1988, an observer invitation was issued to ASOC by the CCAMLR Commission after ASOC provided assurances regarding attendance and confidentiality (CCAMLR 1988). This was not a standing invitation, as it has needed to be reviewed annually (ASOC 1988). However, in practice, ASOC has attended CCAMLR Commission meetings since 1988. They have permission to attend only the plenary, but not the working groups or scientific meetings.

ASOC was invited to participate in the Antarctic Treaty Consultative Meeting (ATCM) on environmental protection in 1990 and has since been granted observer status at these meetings (Darby 1994). This engagement with governments has provided ASOC with a strong support base, and the network is widely respected and influential. ASOC was able to provide knowledge, valuable contacts, support to other groups, and a foundation for campaigning on issues relating to Southern Ocean fisheries.

Two other non-governmental organizations are also invited to participate in the plenary: Coalition of Legal toothfish Operators Inc. (COLTO) and Association of Responsible Krill harvesting companies (ARK). Both are industry coalitions working to promote sustainability to ensure marine living stocks' long-term viability and dependent predators.

COLTO is working to promote sustainable toothfish fishing and fisheries, facilitate its members working together and with others, including through the continued provision of high-quality scientific data to CCAMLR, and provide effective representation for its members. They have been attending CCAMLR since 2004.

ARK has been looking for progress towards stronger cooperation in the entire krill fishing fleet through regular meetings between krill harvesting companies, encouraging the understanding by the fishing industry of CCAMLR's approach and of the scientific requirements for management of the krill fishery, discussion of opportunities and challenges in the krill fishing industry, education and outreach to emphasize CCAMLR's ecosystem approach and the sustainable nature of the fishery, and fostering an agreement from krill fishing companies to support CCAMLR in managing the krill fishery. They have been attending the CCAMLR meetings since 2011.

However, it is different from ICCAT, as NGOs are not allowed to attend the Working Group meetings or the Scientific Committee. They would hardly be considered part of the epistemic community. They are acting on CCAMLR mostly as a transnational advocacy network, which Keck and Sikkink's (1998) definition would be a group of relevant actors working internationally on an issue, who are bound together by shared values, a common discourse, and dense exchanges on information and services. These groups are most prevalent in issue areas characterized by high-value content and informational uncertainty. However, they are not acting on behalf of scientific information, and neither are knowledge-based communities with an authoritative claim to policy-relevant knowledge within their domains of expertise. They are different from epistemic communities, in this case study, because their members do not necessarily share knowledge about the causation of social or physical phenomena in an area for which they have a reputation for competence as well as a common set of normative beliefs about what actions will benefit human welfare in such a domain. They are acting on behalf of a policy agenda.

Conclusion

CCAMLR presents many characteristics that differ from other RFMOs, which would justify the opinion of some stakeholders who are against considering CCAMLR as an RFMO. However, in the current scenario, where stocks are declining, and the RFMOs need to act on behalf of science and the conservation of stocks, it would be better to have the other RFMOs learning with CCAMLR then considering CCAMLR "like a fish out of water."

CCAMLR has been able to implement measures that serve as an example of how we might govern high seas resources more responsively and sustainably. If these approaches are to be duplicated by other RFMOs and expanded to cover more of the world's oceans, it would be illuminating to review the institutional characteristics that might facilitate RFMOs in meeting the conservation and ecosystem-based management directives of the UN agreements.

The institutional design and procedure rules added to the CCAMLR history make CCAMLR a very progressive institution. Knowledge and social learning are considered on a high level, and where truth speaks to power. The epistemic community here has no difficulty being listened to, and after 10 years of documental analysis, their management advice was accepted *ipsis litteris*.

However, if CCAMLR wants to maintain its "uniqueness," then more pro-active measures would need to be taken by CCAMLR members, both individually and collectively. There are, however, several issues (marine protected areas being but one, not explored here in this chapter, but very important), where CCAMLR could, if it so decided, once again demonstrate its international leadership in ocean management with a strong emphasis on conservation.

Chapter 3
ICCAT and the Cooperation for Eastern Bluefin Tuna Management

Introduction

This chapter's focus is the International Commission for the Conservation of Atlantic Tunas (ICCAT) and its international agreement, which is responsible for managing tunas and tuna-like species in the Atlantic Ocean and adjacent seas. ICCAT's responsibility involves fisheries management acting directly by regulating fishing activity through two fundamental processes: allocating the fish among users (countries) and determining the allowable harvest. Both tasks are very complicated and involve political and economic interests. The decisions are embedded in institutional systems; it has to include domestic dimensions. There is growing recognition that the institutional structure, sometimes known as the governance system, and its incentives are the primary determinants of the success or failure of fisheries management. All of this depends on the need of greater skill and knowledge.

As stated on the ICCAT website, "Science underpins the management decisions made by ICCAT".[1] However, in the case of Eastern Bluefin Tuna, the main focus of this chapter, ICCAT, had traditionally set much higher TACs (total allowable catch) than what was recommended by its Scientific Committee (SCRS) (Webster 2011; Sumaila and Huang 2012). With the lack of enforcement and compliance, the Eastern Bluefin Tuna (EBFT) population almost collapsed. Since 2009 there have been changes, and ICCAT began to accept the TAC recommended by the SCRS (ICCAT 2009a, b).

Thus, an analytical question presents itself to be answered in this chapter: when does power listen to science and why? When it does happen, does the fish population show signs of recovery leading to a more effective agreement?

---

[1] http://www.iccat.int/en/.

© The Author(s), under exclusive license to Springer Nature Switzerland AG 2021
L. R. Gonçalves, *Regional Fisheries Management Organizations*,
https://doi.org/10.1007/978-3-030-70362-2_3

Through process tracing (Beach and Pedersen 2013),[2] this chapter analyses ICCAT meeting reports, reports obtained from other international fishery organizations such as the Food and Agriculture Organization of the United Nations, and other peer-reviewed published papers. Also, it includes views of more than 25 such as members of the ICCAT Scientific Committee, Non-governmental Organizations (NGOs), and governmental bureaucrats who attend ICCAT meetings annually to understand the causal connections between knowledge and policy outcomes at ICCAT, particularly in the case study of the Eastern Bluefin tuna.

The ICCAT History

ICCAT's history started years ago, in 1960, when it was recognized that the increase in Bluefin tuna catches was mostly due to the introduction of commercial long liners and purse seiners.

At that time, the Symposium of the Commission for Technical Cooperation in Africa (CTCA) on Tuna in Dakar recommended FAO to convene a conference of plenipotentiaries of all countries interested in the development of the tuna fishery on the high seas and the long-term protection of the resources.

In 1963, at the FAO Council meeting, the rapid growth of tuna fishing in the Atlantic in the absence of coordinated action to study the resources and the effect of fishing upon them was noted. There was a general desire for action to be taken for the conservation and rational exploitation of the tuna resources of the Atlantic. This resulted in the creation of the Working Party on Rational Utilization of Tuna Resources in the Atlantic Ocean, which held its first session at FAO, Rome, 25–30 October 1963, as it was described at the performance review (Hurry et al. 2008) and by Fonteneau (2008).

At the 13th Session of the FAO Conference, the report of the aforementioned Working Party for Rational Utilization of Tuna Resources in the Atlantic Ocean was endorsed, and it was considered that a commission for the conservation of tuna and tuna-like fishes in the Atlantic Ocean was desirable. The Director-General was then authorized to call a conference of plenipotentiaries for the purpose of establishing such a commission and to invite all FAO Member Nations and Associate Members, and all nations, non-Members of FAO that were Members of the United Nations or a Specialized Agency of the United Nations to send duly authorized representatives.

The Conference of Plenipotentiaries on the Conservation of Atlantic Tunas met, on the invitation of the Government of Brazil, in Rio de Janeiro, from 2 to 14 May 1966. The Governments of the following 17 states were represented and signed the final text: Argentina, United States of America, Brazil, Canada, Cuba, Democratic

---

[2] Process tracing can contribute decisively to describing political and social phenomena and evaluating causal claims. It has been used as an essential form of within-case analysis.

Republic of the Congo, France, Japan, Portugal, Republic of Korea, Republic of South Africa, Senegal, Spain, Union of Soviet Socialist Republics, United Kingdom of Great Britain and Northern Ireland, Uruguay, and Venezuela.

On the basis of its deliberations, as recorded, the Conference prepared and opened for signature the International Convention for the Conservation of Atlantic Tunas (ICCAT Convention), which entered into force only in 1969, where it reached the minimum signatures for ratification. ICCAT is, therefore, one of the oldest of the world's five (5) major tuna RFMOs,[3] and it has also become one of the largest.

In the interim period, a meeting of a group of experts in tuna stock assessment was arranged by FAO, under the auspices of the Expert Panel for the Facilitation of Tuna Research, which took place in the Tropical Atlantic Biological Laboratory of the US Bureau of Commercial Fisheries in Miami from 12th–16th August 1968. They concluded that the tuna stocks were heavily reduced by fisheries and that substantive action should be taken for the regulation and management of some of the Atlantic stocks of tunas. However, at the first meeting of the Commission held in December 1969, no management measures were adopted due to uncertain data and insufficient information.

Since the signing of the Convention, the number of contracting parties continued to rise (Fig. 3.1 and Annex II).

Between the ratification process in the 1960s–1970s, as well as the 1980s, the number of countries increased continuously due to a normal process of ratification, which depends on each country but may take more than 10 years. Countries like Angola, Russia, Gabon, Cape-Vert, Uruguay, Sao Tome e Principe, Venezuela, and Guinea Equatorial did ratify the Convention only in later 1980s (see Annex II).

However, after 1994/1995, the number of countries started to once again rise steadily, this time due to two main United Nations remarkable events. The first was

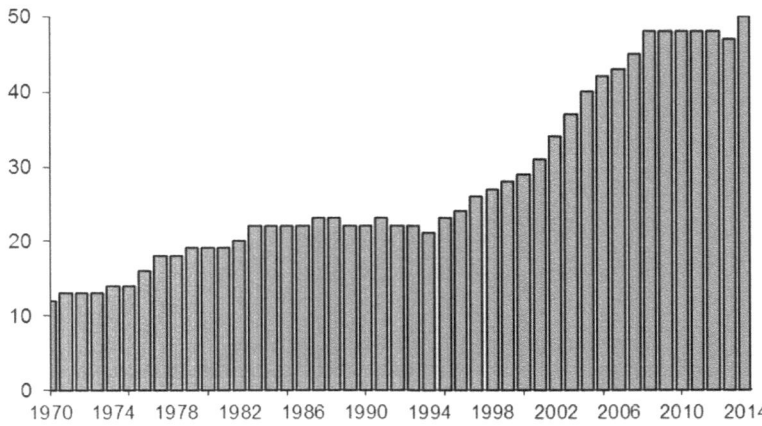

Fig. 3.1 ICCAT contracting parties (*N* = 50). Source: http://www.iccat.int/en/contracting.htm

[3] The other tuna RFMOs are: IATTC (1950), IOTC (1993), CCSBT (1994), and WCPFC (2004).

the entering into force of the UNCLOS (November 1994) itself. The second stemmed from it, and it was framed by another parallel debate: the overarching legal regime for the conservation and management of marine living resources within areas under national jurisdiction and on the high seas, including UNCLOS specific provisions relating to straddling fish stocks and highly migratory fish stocks.

Thus, pursuant to resolution 47/192 of the General Assembly, the United Nations Conference on Straddling Fish Stocks and Highly Migratory Fish Stocks, convened in 1993, completed its work in 1995 with the adoption of the United Nations Agreement for the Implementation of the Provisions of the United Nations Convention on the Law of the Sea of 10 December 1982 relating to the Conservation and Management of Straddling Fish Stocks and Highly Migratory Fish Stocks (the "Agreement"—UNFSA).

The agreement entered into force on 21 December 2001 and currently has 80 States Parties, including the European Union. It is considered to be the most important legally binding global instrument to be adopted for the conservation and management of fishery resources since the adoption of the Convention itself, in 1982. Participation in the Agreement is thus regarded as an important way for a country to signal that it is a responsible fishing nation.

The agreement sets out the legal regime for the conservation and management of straddling and highly migratory fish stocks, with a view to ensuring their long-term conservation and sustainable use. Pursuant to the agreement, the conservation and management of such stocks must be based on the precautionary approach and the best scientific evidence available.

The agreement also elaborates on the fundamental principle established in the Convention that states should cooperate in taking the measures necessary for the conservation of these resources. Under UNFSA, regional fisheries management organizations and arrangements (RFMOs/As) are the primary vehicle for cooperation between costal states and high seas fishing states in the conservation and management of straddling fish stocks and highly migratory fish stocks.

With this, many countries that signed UNFSA, and were not yet involved in any RFMOs, were obliged to sign and be part of an RFMO. The number of ICCAT countries increased by 2010, and is now stable, as most countries that have a fishery interest in the Atlantic Ocean are already part of it.

To add yet another component, in 1995, the FAO Code of Conduct for Responsible Fisheries was also adopted, showing that the international arena was attentive to what was happening after the horizon line.

ICCAT Importance for Fishery Governance

ICCAT was created with the only purpose of managing the fish stocks under its mandate and to maintain their population at levels compatible with the maximum sustainable yield, as provisioned by UNCLOS.

Currently, ICCAT has 50 contracting parties, and a Convention area that covers the entire Atlantic Ocean and the Mediterranean Sea (Fig. 3.1). ICCAT's mandate requires the collection and analysis of statistical information relative to current fishing conditions and population trends carried out by the Standing Committee on Research and Statistics (SCRS) mandate.

Among various responsibilities, ICCAT (1) compiles fishery statistics from its members, cooperating non-members and from all entities fishing for tuna and tuna-like species in the Atlantic Ocean; (2) coordinates research, including stock assessments; (3) develops scientifically based management advice; (4) provides a mechanism for contracting parties to agree on management measures; and (5) produces relevant publications. Contracting parties either have 6 months to implement the management measures adopted or to submit an objection. If an objection is submitted, ICCAT is required to review the objection.

About 30 species are of direct concern to ICCAT in a very large area (Fig. 3.2): Atlantic Bluefin (*Thunnus thynnus thynnus*), skipjack *(Katsuwonus pelamis)*, yellowfin (*Thunnus albacares*), albacore (*Thunnus alalunga*), and bigeye tuna (*Thunnus obesus*); swordfish (*Xiphias gladius*); billfishes such as white marlin (*Tetrapturus*

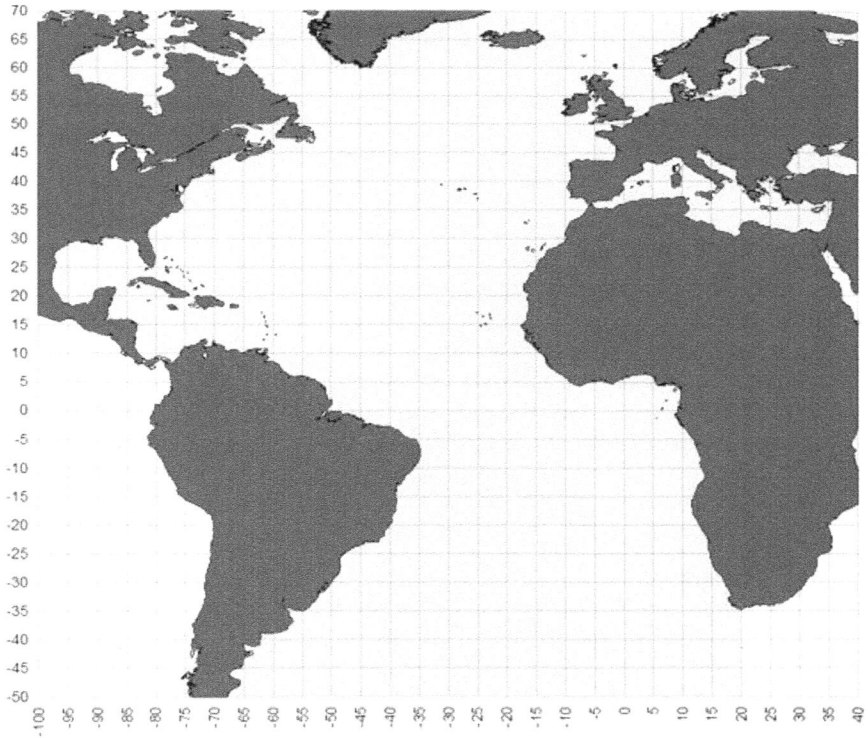

Fig. 3.2 ICCAT Convention area. Source: http://www.iccat.int/en/convarea.htm

albidus), blue marlin (*Makaira nigricans*), sailfish (*Istiophorus albicans*), and spearfish (*Tetrapturus pfluegeri*); mackerels such as spotted Spanish mackerel (*Scomberomorus maculatus*) and king mackerel (*Scomberomorus cavalla*); and small tunas like black skipjack (*Euthynnus alletteratus*), frigate tuna (*Auxis thazard*), Atlantic bonito (*Sarda sarda*), and others.

Not an easy task to manage: many species, different countries, and a great challenge ahead to become a successful RFMO. With the development and entry into force of the United Nations Fish Stocks Agreement (UNFSA) in 1995, the international community made a commitment to strengthen, where needed, Regional Fisheries Management Organizations (RFMOs). Since then, RFMOs have been under increasing pressure to better manage the fisheries resources under their control. The expectations placed on RFMOs have grown over the past decades alongside a proliferation of international hard and soft laws, and there continues to be widespread concern over the performance of RFMOs. This is reflected in calls in international law for organizations such as the United Nations and the FAO to make improvements in the way in which RFMOs operate.

However, a number of RFMOs have undergone significant changes in recent years, with varying degrees of success in terms of ensuring stable cooperative agreements and improved management of the fisheries resources under their control.

In this context, the OECD (Organization for Economic Co-operation and Development) published, in 2009, a study reviewing the experiences of four RFMOs: the Commission for the Conservation of Southern Bluefin Tuna (CCSBT), International Commission for the Conservation of Atlantic Tunas (ICCAT), the North East Atlantic Fisheries Commission (NEAFC), and the North Atlantic Fisheries Organization (NAFO). The objective of the study was to elicit key lessons from the recent experiences of each of these RFMOs in order to inform efforts to strengthen RFMOs, bearing in mind that RFMOs have been, since the past decade, engaged in a process of performance review. The study focuses on the political economic issues underlying the process of implementing change in the structure and operations of RFMOs. It is important to recognize that change occurs both at a large scale (such as major reform and re-writing of a convention underpinning an RFMO) and at smaller scales (such as introducing new catch information systems or dispute resolution mechanisms). The study analyses how the pressure for change arises, how it gains momentum, and how the outcomes are sustained over time. The study also provides insights into ways in which governments and international organizations can help smooth the path of change in strengthening RFMOs.

As it is clearly pointed out in their study (OECD 2009) on ICCAT, the process of change is more difficult due to a relatively large number of contracting parties, a dated Convention, disagreements over scientific assessments, and continued concerns over the overexploitation of key tuna stocks.

However, despite ongoing concerns over the sustainability of particular stocks within ICCAT's responsibility, it is clear that ICCAT has been engaged in a process of changes to strengthen its performance for some years, and this has been showing results, mainly after 2009.

The last decade (2004–2014) has seen a large number of changes focused on improving conservation, management, compliance, and enforcement. Not a small accomplishment, considering those members from NGOs, scientists, governments, and private sector representatives all recognized these improvements.

While there may be some questions over the effectiveness of some of these changes and the extent to which they are actually implemented by some contracting parties to ICCAT, the changes have helped to move the organization towards a more effective framework. This is reflected in the success management stories of some specific stocks under ICCAT management, such as the recovery of the Atlantic swordfish stocks, illustrated in the OECD (2009) study.

Another very emblematic and widely discussed case under ICCAT management that is under improvement is the Eastern Bluefin tuna. The topic is so important that it was considered crucial, and the performance review stated that the judgment of the international community on ICCAT would be based largely on how the management of fisheries on Bluefin tuna (EBFT) would be accomplished along the years. Due this importance, the EBFT was chosen to be the case study of this chapter.

Eastern Bluefin Tuna Case: A Lesson to Learn

The Atlantic Bluefin tuna, native to both the western and eastern Atlantic Ocean, may be naturally divided into two stocks: western and eastern Atlantic BFT, which differ both in their habitat and their life histories. Both groups of BFT are highly migratory and have a long life span of up to 30 years. The eastern Atlantic BFT stock is targeted by different types of fishing gear (longline, purse seine, traps) in the eastern Atlantic Ocean and the Mediterranean Sea.

Fishing Bluefin tuna has a very long tradition in the Mediterranean Sea. The first evidence is estimated at around 7000 BC (Fromentin and Ravier 2005). The popularity of Japanese sushi and sashimi worldwide during the 1980s made the BFT much more attractive economically than ever before. Consequently, vessel capacity, vessel power, and new storage innovations for BFT experienced tremendous increases in the 1980s and 1990s, which imposed severe pressure on the EBFT stock (Sumaila and Huang 2012).

The duty to set and allocate Eastern Bluefin Tuna's (EBFT) catch quotas according to its scientific stock assessment is one of ICCAT's major responsibilities. In the past, not long ago, ICCAT had consistently set the quotas much higher than the levels recommended by its scientists (see Fig. 3.3). Furthermore, compliance and enforcement associated with those quotas were weak, which led to a huge mismanagement problem (Mackenzie et al. 2009; Sumaila and Huang 2012).

From 2007 to 2009, a series of events gave prominence and attention to the crisis of BFT fishery and exposed the lack of political action and management of fish stocks. The ICCAT contracting parties, cooperating non-contracting parties, and entities or fishing entities (CPCs, hereafter) were feeling the pressure and the need for change in the way decisions were made at ICCAT to avoid the BFT crisis. Not

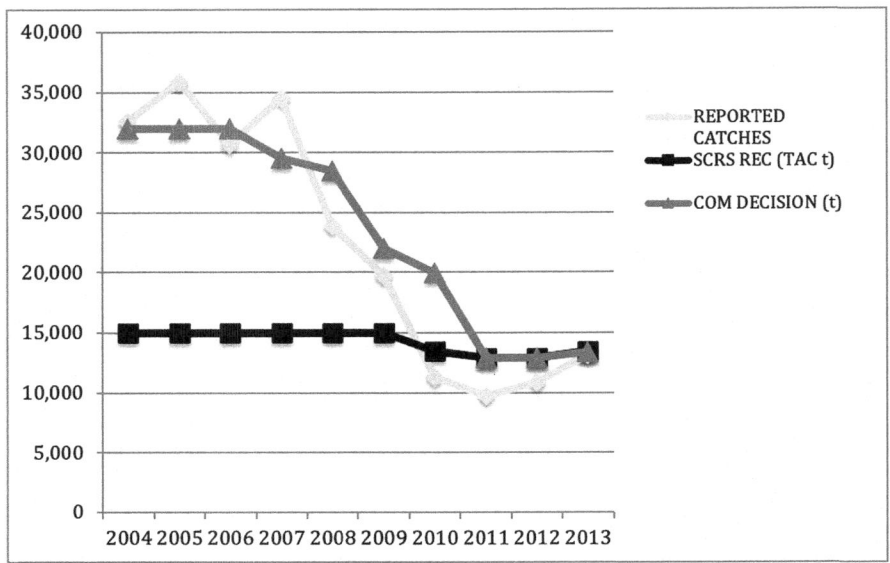

Fig. 3.3 Reported catches, SCRS TAC recommended and the Commission decision. Source: Author

only due to environmental concerns, but also because pressure was coming from various sectors and stakeholders.

Diverse events were happening to expose the situation. In 2006, Canada in its opening statement warned on the "need to agree upon a set of measures that will strengthen compliance and allow us to establish a solid rebuilding plan for East Atlantic Bluefin. Without this basic step, we will run the risk of having others, such as CITES, step in and do what we cannot or will not" (ICCAT 2006a, p. 82).

On the following year, 2007, Japan also mentioned a concern that "as the current plight continues, Appendix II, or even I for Atlantic Bluefin is a likely eventual action by CITES in 2010. That is a cessation of commercial Bluefin fisheries in the entire Atlantic" (ICCAT 2007a, p. 73).

However, 2009 was a great turning point when the fishery crisis became even more evident. The SCRS report warned about the threat of EBFT stock to collapse and it highlighted that the uncertainty of under-reported catches was playing a major role (ICCAT 2009b, p. 45).

Not only was the SCRS ringing the bell, but non-SCRS scientists were also warning that the EBFT stock had been brought to near collapse as a result of an increased fishing effort, high demand of tuna at the markets, and lack of effective management intervention (Webster 2009; Mackenzie et al. 2009).

As pointed out by Adler and Haas (1992), crises and new developments not only accelerate the diffusion process but also lend urgency to the task of reevaluating current policies and coming up with alternatives. Crises and uncertainty triggered a search for expertise illustrating an exact situation where social learning may occur.

A crisis is in place warning that due to overexploitation and mismanagement a fish population may collapse, so states recognize that they need to deal with the problem and they delegate the task of research and providing information to the experts.

Thus, bearing this entire background context, ICCAT began to cut the TAC (total allowable catch) substantially, from 32,000 tons in 2006 to 12,900 tons in 2011 and 2012 (Fig. 3.3). The warning sent from the SCRS in 2009 pointed to the fishery crisis and triggered this action. However, the acceptance of the SCRS recommendation does not entirely mean that all the problems were solved. ICCAT was still failing on compliance and enforcement and countries were still failing to provide accurate fishery data, which resulted in poor and underreported fishery information. Therefore, at that time, to recognize, demand, and accept the scientific advice was a big step.

In 2014, the SCRS report showed that "the implementation of recent regulations has clearly resulted in reductions in catch and fishing mortality rates. All CPUE[4] index show increasing trends in the most recent years" (ICCAT 2014).

Additionally, the 2015 Report of the Standing Committee on Research and Statistics (ICCAT 2015) shows that the decisions taken, based on SCRS advice, made an impact reducing the fishing mortality (F 2.5 and F 10.) and the strong recovery of the spawning stock biomass (SSB) (Fig. 3.4).

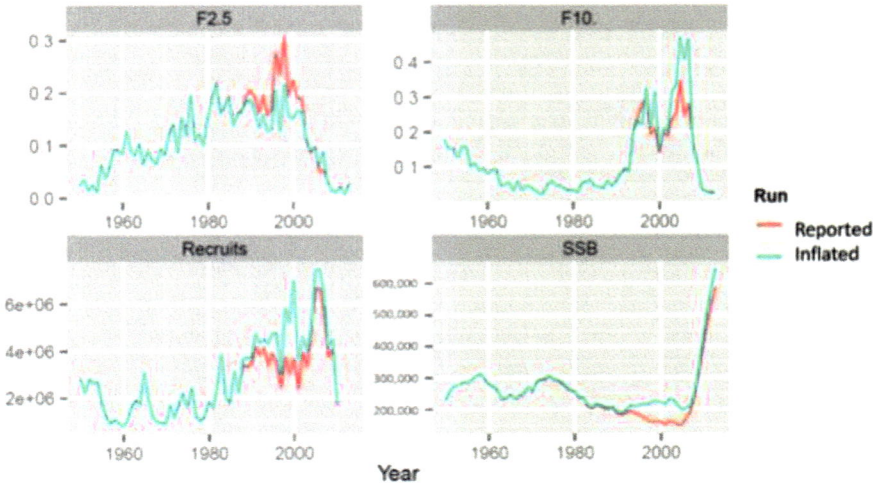

Fig. 3.4 Fishing mortality (for ages 2 to 5 and 10+), spawning stock biomass (in metric ton), and recruitment (in number of fish) estimates. Red line: reported catch; blue line: inflated (from 1998 to 2007) catch. Source ICCAT 2015, page 108

---

[4] In fisheries and conservation biology, the catch per unit effort (CPUE) is an indirect measure of the abundance of a target species. Changes in the catch per unit effort are inferred to signify changes to the target species' true abundance. A decreasing CPUE indicates overexploitation, while an unchanging CPUE indicates sustainable harvesting.

The Committee listened to science, and the EBF tuna populations started to show signs of recovery. Still a big empirical question remains open. Why ICCAT has begun to accept the TAC recommended by the SCRS only since 2009?

The Emergence of an Epistemic Community

This chapter argues that, under the widely publicized crisis about the international tuna fishery management, an epistemic community emerged under ICCAT to provide new causal arguments that enabled the CPCs to make sense of the situation, and to employ collective efforts to deal with shared problems.

As indicated by Haas (2014), epistemic community ideas contribute to causal mechanisms of social learning, which lead to more scientifically informed, and often comprehensive, approaches to policy-making and dealing with issues.

Epistemic communities have been conceptualized as a group of experts who persuade others with basis on their professional knowledge. To be part of this knowledge community, the individuals must have the expertise necessary to understand the issues at stake, to interpret the information similarly, and then to share the same goals about what should be done. The group's policy aims have to reflect their expert knowledge—and not some other motivation—otherwise they lose authority and legitimacy with their target audience, which in the area of international fishery are usually the CPCs representatives. In other words, epistemic communities must have an authoritative claim on knowledge to impact policy outcomes.

This chapter claims that an epistemic community emerged under the SCRS, and consisted of the ICCAT chairman, some NGOs, individuals, and some national scientists.

The science and most of the knowledge used to formulate policies at ICCAT are developed under its Scientific Committee of Research and Statistic (SCRS). The SCRS is built with a high caliber of scientists indicated by CPCs. They are responsible for developing and recommending to the Commission all policy and procedures for the collection, compilation, analysis, and dissemination of fishery statistics. It is the SCRS' task to ensure that the Commission has available at all times the most complete and current statistics concerning fishing activities in the Convention area as well as biological information on the stocks that are fished. The SCRS also coordinates various national research activities, develops plans for special international cooperative research programs, carries out stock assessments, and advises the Commission on the need for specific conservation and management measures.

They have a clear mandate for providing advice to the Commission, and as noted from the interviews, they do it with a very high level of professionalism. Along the period observed here (2004–2013), 924 scientists attended the SCRS meetings in total, with a 92.4 participants-per-year average (Fig. 3.5). Even when those scientists are working for the government and with the probable influence of politics on their work, they are professionals with academic backgrounds and are doing their best to keep their analysis on biological data, as it is possible to note from the

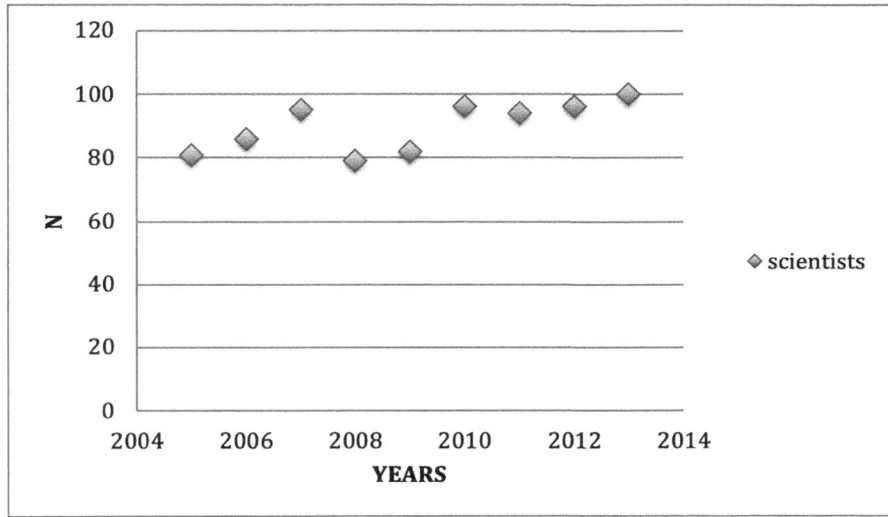

Fig. 3.5 Number of scientists attending the SCRS meetings along the study period

scientific reports and data analysis. Some of them, more than others, but in general, the information they are producing suffers more from the lack of accurate data and from the underreported information from the CPCs, than from the politicization of science on the data analysis (see discussion at first chapter about the institutional design of ICCAT).

Not all national scientists would qualify as epistemic community members. Through snowball interview technique, research on peer-reviewed papers, and understanding their positions during the meetings and media reports, it was possible to find some specific names ($N = 34$) that would respect the criteria established by Haas (2004b) to be members of its epistemic community (Annex III). They are the ones whose policy aims were reflecting their expert knowledge and no other motivation (Cross 2015). Those national scientists were stretching their boundaries on providing advice to the Commission and were persuading decision-makers to heed their knowledge. Also, they were qualified by their professionalism, meeting frequently (those were in more than 80% of the SCRS meetings), in different meetings more than twice a year, and they are trained on the use of the same methodology. They also know who shares the same beliefs and common values among themselves. This was made possible by capturing the interviews of 25 stakeholders that attended ICCAT.

Beside the scientists, observers are also allowed to participate actively in SCRS meetings. Prior to the formal recognition of the observer status in 1999, scientists from NGOs or industry consultants participated in SCRS activities as members of national delegations. The option to do so continues to exist and some still do participate as members of their national delegations rather than as observers.

As observers, they are allowed "to participate in all meetings of the organization and its subsidiary bodies, except extraordinary meetings held in executive sessions or meetings of Heads of Delegations".[5] Any eligible NGO admitted to a meeting may: attend meetings, as set forth above, but may not vote; make oral statements during the meeting upon the invitation of the presiding officer; distribute documents at meetings through the secretariat; and engage in other activities, as appropriate and as approved by the presiding officer.

There are some specific NGOs that are involved in the plenary meeting, but are also investing time and dedication to contribute to scientific meetings as well. The profile of NGOs that also attend the SCRS meetings as observers, however, differs from those who normally attend the plenary only. According to the attendee list for the study period, 51 NGOs attended the SCRS meetings; however, only 4 of them were deeply involved in the scientific work, actively contributing to the EBFT discussion, being present in more than 50% of them. Their staff, normally not a large number of people, possesses an academic background on science that would enable them to contribute to the fishery science debate (Table 3.1). Those organizations, and also those specific individuals, were frequently mentioned by the national scientists, decision-makers and by their peers in the interviews with respect, credibility, and recognition of their contribution to the scientific debate. They were included in the epistemic group for the EBFT, who shared the same beliefs, shared causal beliefs, or shared professional judgment with others while also presenting common notions of validity and common policy enterprise in terms of the subject.[6]

Another point that must be noted is that some of the scientists that are part of SCRS also attend the plenary meeting which would mean that they may have access to decision-makers, NGOs, and other stakeholders. The SCRS members may also

Table 3.1 NGOs representatives at ICCAT

Organizations	Individual members
OCEANA	Cornax, María José
WWF	Tudela, Sergi
Federation of Maltese Aquaculture Producers	Deguara, Simeon
ISSF	Restrepo, Victor

[5]According to RES 05-12—Guidelines and criteria for granting Observer status at ICCAT meetings (www.iccat.int)

[6]During the entire study period, Bird Life International have also attended all SCRS meetings, represented by Dra. Cleo Small; however, she was not involved in the Bluefin tuna debate, but on the bycatch and Sea Birds topic, reasons for why she is not considered here as part of the epistemic community. No doubt, they are part of an ICCAT epistemic community.

be very influential, and even be elected as a Commission Chairman. This was the case of Mr. Fabio Hazin, ICCAT chairman from 2007–2011.[7]

Mr. Hazin is certainly part of the epistemic community and his professionalism and leadership in this case were also a factor that needs to be considered, as it was mentioned by many others during the interview process. As a Ph.D. fishery scientist himself, all of his political statements were based on the scientific information provided by the SCRS. Also, as he used to be part of the SCRS, he knew NGOs, and scientists by a long period of time for whom he expressed a trust. Over the years, he also created a strong alliance with key and very active researchers, such as Gerald Scott (USA) and Josu Santiago (UE), both SCRS chairmen and also part of the epistemic community, what undoubtedly contributed in persuading the CPCs to make their decisions based on science. With all the knowledge he had, and with the reliable relationship with many CPCs commissioners, as a chairman, Mr. Hazin was in a high-level position at the negotiation table with access to multiple key decision-makers, that he could persuade to follow the scientific advice, which he knew, for a fact, was the right thing to do.

At the soonest opportunity he took over the chairman position for the second mandate in 2009 when he clearly stated: "let's not fool ourselves: there will be no future for ICCAT if we do not fully respect and abide by the scientific advice. If we do not follow the instructions science is giving us, our credibility will be irreversibly jeopardized and the mandate to manage tuna stocks will be surely taken out of our hands" (ICCAT 2009a, p. 72). During his 4 years as Chairman, science was his guide for BFT and other species. He is recognized by his peers from SCRS as part of the group, even while he was playing a role as a Commission chairman.

The year of 2009 was definitely not only a turning point, but it was the impetus for the momentum wherein science began to guide the actions related to EBFT. As Mr. Hazin mentioned in the opening statement of 2010, "differently from the previous two years, this year I feel that I no longer have to emphasize the need for ICCAT to follow the scientific advice, not because this is not important anymore, but, on the contrary, because in my view the obligation to respect science has finally become firmly entrenched in the work of this Commission" (ICCAT 2010a, p. 41).

Particularly in 2009, but in the years following, Mr. Hazin performed his leadership using two main techniques in persuading key states to use science as a guide, and creating an alliance between key and powerful countries, such as the United States, Canada, and Brazil, to persuade Japan and the European Union to follow scientific advice—which at the end was a very successful strategy.

As a person and as chairman, and still as a part of this epistemic community, he could support scientific knowledge and take that knowledge to the decision-makers.

---

[7] When Dr. Fabio Hazin took over the ICCAT chairman position, he had been attending ICCAT meetings since 1998, representing the Brazilian government as well as working as a scientist at SCRS. For the entire period of this study, which also coincides with the period where the Commission accepted the TAC proposed by the scientists, he has not missed any plenary or SCRS meeting. He also has been involved in other international organizations, such as CITES and FAO.

He was in a position that allowed him to influence the decision-makers, as he had open access and their reliability as well.

Following the definitions established by Haas (1992, 2014) and also Cross (2013), this chapter argues that there was an epistemic community influencing policy decision at ICCAT for the EBFT and it contained scientists from different groups. Some of them were scientists that represented the CPCs, others were scientists from specific NGOs, and also the chairman from the SCRS and from the Commission, as they were all part of the network of professionals that possessed the necessary expertise, interpreted the information and data with similarity, and were willing to persuade others based on their knowledge.

According to Cross (2013), they would also match the epistemic community criteria as they go beyond formal expectations as a group. As the attendees list shows, they also met annually, holding the same position at various times, developing *esprit de corps*, and sharing distinctive cultures and professional norms beyond the bureaucracy as they spent more time on technical issues than formal bureaucracies. Adding to that, using snowball technique interviews, an interesting point to be highlighted is how easy it was for the members of SRCS to identify their allies, the ones with whom they shared the same beliefs, independent from the role or the organization they were representing at that moment, as it changed.

Most of the epistemic community members in this case, when asked if they were part of NGOs, government or academia, answered they had been part of all of them. They have played with different hats, but always using science to persuade decision-makers using their common knowledge as a guide.

The differential for this epistemic community is that, as it was constituted by different stakeholders, it was able to influence the decision-makers from different angles providing a greater transmission of knowledge. The advantage of its diverse constitution was that they had similar professional academic backgrounds and many different open channels to persuade decision-makers, as each of them had their own strategies and their own areas of influence.

As the SCRS has a very clear mandate to provide information and advice for the management of tunas, not all scientists are comfortable in a position to persuade decision-makers regarding the need to accept scientific recommendations as policy decisions. Therefore, as a strong epistemic community, it sought to go beyond their formal professional role as a group, and it was often able to persuade CPCs to fundamentally change the policy decision. In this case, the epistemic community influenced the decision-makers, but also transmitted knowledge through NGOs and international organizations like CITES.

Usable Knowledge for EBFT

As Haas and Stevens (2011) claim, in order to influence policy, an epistemic community needs to produce usable knowledge. Usable knowledge, borrowing their definition, is "a substantive core that makes it usable for policymakers and a procedural dimension that provides a mechanism for its transmission from the scientific community to the policy world, and it allows for agency when theorizing about broader patterns of social learning, policymaking, and international relations."

The general factors that influence the likelihood that decision-makers will apply scientific knowledge are credibility, legitimacy, and saliency (Schroeder et al. 2008). Credibility means that the knowledge claims are believed to be accurate: within a consensus theory of truth, this means that they are publicly created through a deliberative process by people widely regarded as experts. Legitimacy means that knowledge is developed by people who have social authority and that it is accepted by people outside the community that developed it. In practice, the assignment of legitimacy often rests on peer review and scholarly reputation. Salience means that the information is timely and is organized on a politically meaningful timescale and scale of resolution.

The knowledge produced by ICCAT's epistemic community is accurate, and transmit credibility and legitimacy for most of its members. Even independent scientists agree that the information is correct and that the main problem is political due to the rejection of scientific advice (Mackenzie et al. 2009; Sumaila and Huang 2012).

The epistemic community group is constituted by high-level scientists, most of them with a Ph.D. or with research focusing on fishery science or fishery management. As highlighted by the performance review, the analyses used by the SCRS to formulate advice are prepared by CPC's scientists from government agencies, universities, NGOs, and consultants, and they are peer-reviewed through a rigorous three-stage process. The structure of the process, the diversity of participants/analysts, and the large number of people involved do not guarantee that errors will not be made, but it provides a reasonable assurance that if errors are made, they will be discovered, admitted, and corrected. In addition, from 2004 to 2013, of the 331 scientists who attended the SCRS meetings, about 62 were present at more than 50% of them, which guarantee that, with time, the SCRS can now count on a larger number of scientists who are trained and competent in fishery methods and analysis. No doubt that the information presented by the SCRS is the best estimate they can have with the data they have at hand, being thus the "best scientific advice available" requested under Article 61 of the United Nations Convention on the Law of the Sea.

The greatest impediments to the completion of more reliable stock assessments are the lack of data and incomplete knowledge. The lack of reliable data on catches for important components of the fisheries forces scientists to make assumptions and guesses on the amounts caught and catch rates, which have the potential to increase substantially the uncertainties in the assessments.

All the knowledge is revised and methods and data are improved every year. The knowledge has saliency since it is discussed 1 month before the Commission plenary meeting, when it is then presented for political discussion. The SCRS chairman makes the presentation at the plenary in a very clear summary that contains all of the technical and scientific information and the whole report. A very comprehensible chart is also circulated before the meeting as supporting information for decision-makers.

According to Haas (2004a, b), scientific consensus can inform policy when groups responsible for articulating consensus have stable access to decision-makers. However, for consensus to be acceptable to leaders it must emerge through channels that are viewed as legitimate by the leaders, just as the role performed by the epistemic community was done here. Typically, this happens when the scientists have a reputation for expertise, when the knowledge was generated beyond suspicion of policy bias by sponsors, and when the information is transmitted to governments through personal networks (Haas 2001).

The Role of NGOs on the Eastern Bluefin Tuna Debate

NGOs participate in global environmental politics in a number of ways: they try to raise public awareness of environmental issues; they lobby state decision-makers hoping to affect domestic and foreign policies related to the environment; they coordinate boycotts in efforts to alter corporate practices harmful to nature; they participate in international environmental negotiations; and they help monitor and implement international agreements (Betsill and Corell 2001). At ICCAT, they not only do all of the above, but since some of their staff (Table 3.1) are willing to contribute to the scientific debate, as scientists, they also qualify to be part of the epistemic community.

There were 45 NGOs attending ICCAT meetings as observers during this research period. This group contains international environmental organizations (e.g., Birdlife International, The Pew Environmental Group, Greenpeace, and World Wide Fund for Nature (WWF)); and includes associations that advocate for fisheries and fishermen (e.g., International Seafood Sustainability Foundation (ISSF), Tuna Producer Association, Association Euroméditerranénne des Pecheurs Professionnels de Thon- AEPPT), and also Universities (Annex IV).

Observing their frequency both at Commission and at SCRS meetings, it is possible to notice that most NGOs mainly attend the Commission meeting. Seeking to achieve their goals under limited human and financial resources, most of them routinely and strategically opt to do their work within the political meeting (Fig. 3.6).

NGOs are a key group at ICCAT and throughout time they have gained more space and a stronger voice in the negotiations. International NGOs and transnational advocacy networks have specific policy goals that are based on shared causal beliefs about what actions will result in the achievement of their aims. However, they are different from the epistemic community described above. "Their goals typically derive from idealist interests such as environmental protection and social change. They have some specific characteristics: the centrality of values or principle ideas, the belief that individuals can make a difference, the creative use of information, and the employment of non-governmental actors of sophisticated political strategies

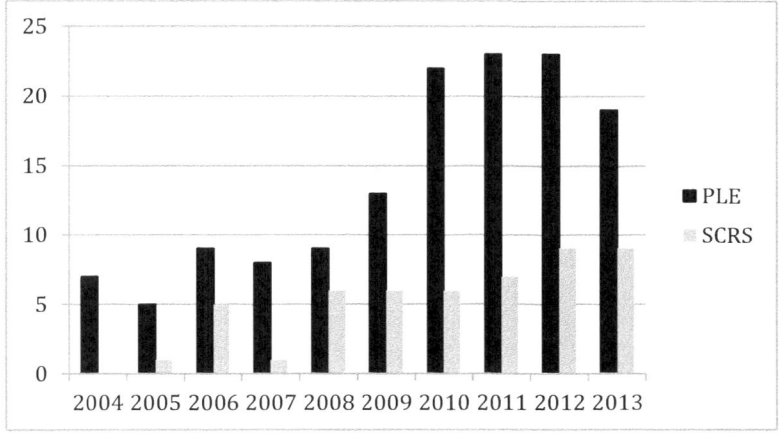

Fig. 3.6 Number of NGOs attending Commission meetings and the SCRS meetings during the study period (2004–2013)

in targeting their campaigns" (Keck and Sikkink 1998). They are value-driven and thus seek to change both policy outcomes and the terms of the debate (Cross 2013).

In this EBFT case, most of them would not be considered part of an epistemic community, as they were not necessarily acting on behalf of knowledge, with the exception of the few mentioned above (Table 1) who were frequently attending SCRS meetings. Therefore, they supported the epistemic community in building more consensual knowledge and to disseminate it to civil society and to the media, helping to enhance the awareness and the importance of decision-makers listening to science.[8] From 2009 to 2013, in most of their opening statements, observers mentioned the need to follow the SCRS' advice and to use the best scientific information available. Their participation at the Commission meeting and their involvement at the ICCAT EBT debate helped to bring attention to the topic and to put pressure on the CPCs to listen to science.

Their statements to media were supporting science as well. "We encourage policy makers to continue to listen to science in the future. Only then will the East Atlantic and Mediterranean stock of Atlantic Bluefin tuna have a chance to fully recover," said Dr. Sergi Tudela, Head of Fisheries, WWF Mediterranean. Another NGO also supported the decision along the same line: "It is encouraging that ICCAT listened to the recommendations of its own scientists and agreed to keep catch limits for bluefin tuna within their advice. This decision will give this depleted species a fighting chance to continue on the path to recovery after decades of overfishing and

---

[8] Link to NGOs press release: http://www.greenpeace.org/international/en/press/releases/greenpeace-cites-is-last-chan/.

mismanagement," says Susan Lieberman, international policy director, Pew Environment Group. They were acting on behalf of science, backing up epistemic community knowledge.

It is also noteworthy that 2009 to 2013, when the CPCs reduced the TAC for EBFT, were the years when a larger number of NGOs were attending ICCAT meetings (Fig. 3.6), which suggests that the topic was gaining greater media coverage and also a better understanding by public opinion. The transnational advocacy network supported the epistemic community and this helped greatly to increase pressure on decision-makers at international and domestic levels. Most of the stakeholders interviewed assumed that the NGO's pressure helped to expose the ICCAT EBFT mismanagement and created an international embarrassment which CPCs wanted to avoid.

The relationship between NGOs and SCRS was not always a relationship based on trust, however. During the interviews, NGOs, like Greenpeace, were blaming the SCRS for the high interference from politics on science, and clamoring for a peer review of their results. However, even with the criticism, from 2009 to 2013, most of them were speaking in one unique voice for ICCAT to follow the SCRS advice and reduce the Bluefin tuna TAC. This created greater pressure on CPCs and helped the decision-makers to accept science. Another factor that helped NGOs to increase the SCRS' knowledge was that since 2004 the number of NGOs attending the SCRS meeting increased (Fig. 3.6), bringing more legitimacy, transparency, and reliability between the civil society and the epistemic community.

Those NGOs that attended the SCRS meetings were essential in this process; they contributed scientific information and, in the end, were part of the epistemic community. But they also did advocacy and policy work and sponsored the science that had been done by the SCRS. By speaking in a unique voice, they transmitted a consensual and usable knowledge to the decision-makers and also to the civil

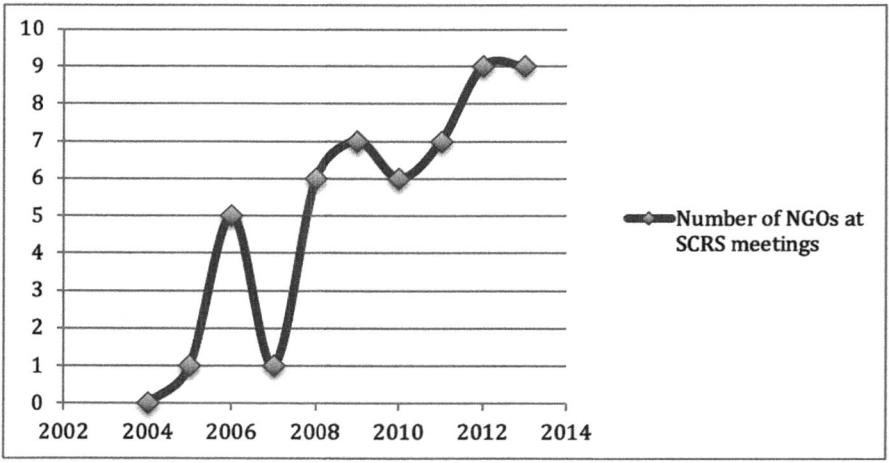

Fig. 3.7 Number of NGOs attending the SCRS meetings

society. All of this together created an environment of persuasion that was felt by the decision-makers, as it was also mentioned by the interviewed stakeholders (Fig. 3.7).

As some NGOs were involved in SCRS meetings and also in Commission Meetings, it seems they were able to create a bridge between the SCRS and NGOs, which made the information even more powerful and trustworthy.

Normally, in IR, the science-politics interface has been framed primarily as a matter for scientists and decision-makers. Scientists inform policy-makers and policy-makers turn to science for knowledge and technical assistance. This case, as it was argued by Bäckstrand (2003), suggests a triangular interaction between scientific experts, policy-makers, and citizens, where citizens are involved and aware of the problem through the information disseminated by the NGOs and the media. Thus, the citizens were not just a recipient of policy but also an actor in the science-policy nexus, influencing and exerting pressure at the domestic level.

The Threat of CITES

The proposal to include the Bluefin tuna (the West Atlantic and the East Atlantic stocks) into the Appendix I of the Washington Convention (better known as CITES) was promoted by the Government of Monaco in July 2009 and subsequently endorsed by various countries. The Appendix I lists species that are the most endangered among CITES - see Article II, paragraph 1 of the Convention). As Webster (2009) highlights, this action had a precedent. They considered listing Bluefin tuna as an endangered species almost 20 years ago. In 1991, Sweden nominated the western stock of Atlantic Bluefin tuna just as Monaco nominated the eastern stock in 2009.

The 2009/2010 proposal was mostly based on science attributes such as the low level of the Bluefin tuna population, the unsustainable level of fishing effort, the outputs of the assessment provided by SCRS with the following low recovery figure of the population in the wild, and the mismanagement of this fish resource by ICCAT and all countries concerned.

The proposal text makes reference mostly to SCRS documents and WWF reports showing a clear influence of the epistemic community members on the construction of the Monaco proposal.[9]

During the 2009 ICCAT meeting, the possibility of Bluefin tuna being listed in the CITES Appendix 1 was so great that most countries, during their opening statements, mentioned this fact and the desire to work for improving the performance of ICCAT in order to continue to be the sole organization responsible for BFT management. The impact of CITES was powerful and the threat was strongly felt by the CPCs, mainly Japan.

---

[9] http://www.cites.org/eng/cop/15/doc/E15-52.pdf.

A strong example was the statement made by Brazil, a key player in 2009 as a host for the meeting: "ICCAT is now facing the risk of losing the mandate to manage the bluefin tuna stock, mainly because it has failed to abide by the scientific advice. It is needless to say how such a development could jeopardize the future of this Commission. In light of that, we reiterate the plea we made last year for all Contracting Parties to embrace the cause of leaving the meeting in Brazil with all measures adopted by the Commission in full conformity with scientific advice, not only in relation to bluefin tuna, but to all species under the mandate of the Commission" (ICCAT 2009a, p. 74).

Japan, who was concerned about CITES, and who voted against the proposal at the CITES meeting, also made a commitment in 2009 "to work out at this meeting conservation programs consistent with SCRS advice for not only bluefin tuna but all major species" (ICCAT 2009a, p. 77).

Additionally, the United States has argued "Now, however, the global scrutiny on ICCAT has intensified, particularly in light of the recent proposal to list Atlantic Bluefin Tuna on Appendix I of the Convention on International Trade of Endangered Species of Wild Fauna and Flora (CITES) (ICCAT 2009a, p. 80)."

In the end, Monaco's proposal at CITES was not approved in 2010, but, as many interviewed stakeholders mentioned, the pressure was felt by the CPCs. Some of them would say that during this 10-year period, the serious attitude and strong will of ICCAT to improve its conservation efforts for this stock overweighed this CITES challenges and that ICCAT had started its work to rectify the situation of EBFT well before CITES took up this issue. However, according to WEBSTER (2011), "with the threat of a CITES listing looming in 2009, the Commission adopted catch limits that would quickly reduce legal harvests below the scientifically recommended levels."

However, based on the interviews with key stakeholders and meeting reports, this chapter states that CITES played an essential role in creating pressure on the CPCs and exposing the crisis within ICCAT management. However, the CPCs did not agree to lower the quotas due to the threat from CITES, since ICCAT had started before 2010 to rectify the situation, well before CITES took up this issue. An important sign of this was when the Commission asked for advice from the SCRS (ICCAT 2009b, p. 45) and established a recovery plan.[10]

The ICCAT listened to the epistemic community arguments because of the nature of the way the science was organized, their usable knowledge based on real information, and the reliability relationship created between them, enhancing their legitimacy and credibility. Even CITES was influenced by the epistemic community knowledge when it made the proposal for listing Bluefin tuna[11] (Fig. 3.8).

In stating that, this chapter is not reducing the role of NGOs or CITES to a mere megaphone of scientific information generated by the epistemic community, as

---

[10] The 2010 TAC was revised to 13,500 t by [Rec. 09-06], which also established a framework to set future (2011 and beyond) TACs at levels sufficient to rebuild the stock to BMSY by 2022 with at least 60% probability.

[11] http://www.cites.org/eng/cop/15/doc/E15-52.pdf.

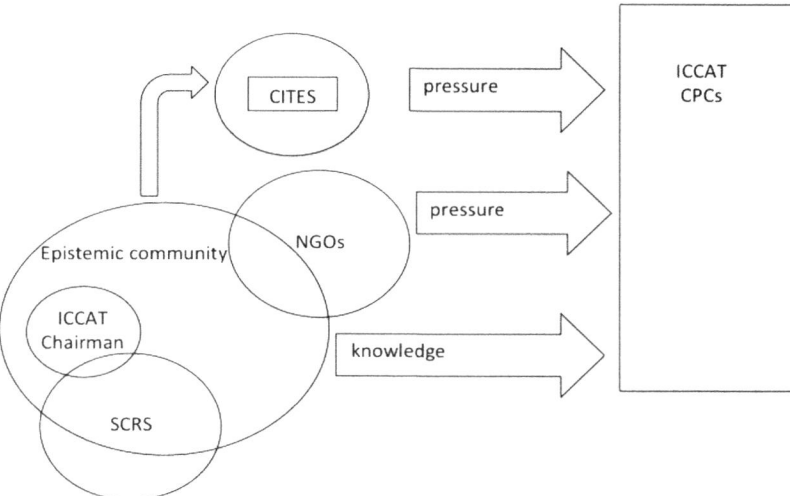

Fig. 3.8 Epistemic community formed by ICCAT chairman, part of SCRS national scientists, and part of NGOs. They were consistently providing knowledge to ICCAT CPCs, while CITES and NGOs were creating the sense of urgency

Toke (1999) could argue. On the contrary, NGOs and CITES had a very protagonist role, backing up the science and the epistemic community knowledge and advocating in a unique and consensual voice to keep up the tune and the need for change. Sponsoring the science did not mean they did not have their own agenda, it only meant that the agents were acting on behalf of a legitimate knowledge, provided by a powerful epistemic community, to shape policy decisions and to reach joint solutions.

Conclusion

The Bluefin tuna case demonstrates that when power listens to science and implements that advice through a social learning process, fish stocks can recover and thus reach the main goal of the Commission, conferring more effectiveness for the international agreement. Power's willingness to listen to science was one of the first steps towards improving ICCAT's effectiveness. However, CPCs still need to improve the quality of data provided to the SCRS and ICCAT still needs to address the lack of compliance and enforcement, which are reducing the chances of promoting proper management.

The EBFT case illustrates a situation where a fishery crisis was in place, in an environment full of uncertainty, and when a transnational network of NGOs and

CITES helped to enhance the pressure on the decision-makers to take action by that time the epistemic community was ready to provide usable knowledge to the CPCs.

The epistemic community was formed by a group with a high level of professionalism; it emerged from the SCRS and included scientists from NGOs, CPCs, and the ICCAT chairman, a scientist himself, producing usable knowledge that was coherent, legitimate, and solid enough to influence strategically within many dimensions of power. The knowledge produced about the state of the EBFT population was sound enough to have other key stakeholders supporting the epistemic community claim.

This chapter is a contribution to knowledge and it opens a path towards a clearer understanding of how social learning can change policy decisions at RFMOs. The history of EBFT management at ICCAT is a very emblematic case in international cooperation where an epistemic community spoke loudly to power, and power listened to them. This influence was only possible due to the fact that NGOs and CITES reported the emergent fishery crisis that could damage ICCAT history. Additionally, they created pressure on ICCAT by sponsoring the epistemic community knowledge.

With its analysis, the chapter contributes to recent efforts to arrive at a more nuanced theoretical and empirical understanding of effective governance mechanisms at the transnational level. A future study on the other case studies within RFMOs to find other conditions which force power to listen to science would be beneficial.

To conclude, the study of epistemic community and the science and policy interplay to discuss international cooperation do not exclude the central importance of interstate relations in world affairs. However, to not take into account the non-state actors and the role of knowledge in today's world leaves one with only a partial picture of the international system, and thus might represent an incomplete understanding of world politics itself.

Chapter 4
CCSBT and the Management of Southern Bluefin Tuna

Introduction

Southern Bluefin tuna (SBT) were heavily fished in the past, with annual catches reaching 80,000 t in the early 1960s. Heavy fishing resulted in a significant decline in the numbers of mature fish and the annual catch began to fall rapidly. By the mid-1980s, it was apparent that the SBT stock was at a level where management and conservation were required and there was a need for a mechanism to limit catches. From 1985, the main nations fishing for SBT at that time, Australia, Japan, and New Zealand voluntarily agreed to apply strict quotas to their fishing fleets to enable the rebuilding of the stock.

In 1994, these arrangements were formalized with the signing of the Convention and the establishment of the Commission for the Conservation of Southern Bluefin Tuna (CCSBT)—one of the newest RFMOs. This international organization was founded by Australia, Japan, and New Zealand; and in 2002, Korea and Taiwan joined.

The CCSBT manages the SBT fisheries by setting a global total allowable catch (TAC) and the fishing quotas for each country, but has often failed to reach agreement on catch limits (Kurota et al. 2010). The Scientific Committee (SC), aided by its internal Stock Assessment Group (SAG), makes management recommendations to the Commission. However, debate between members over stock status due to large uncertainties in stock assessment results prevents management action.

In this context, through process tracing (Beach and Pedersen 2013),[1] this chapter analyzes CCSBT meeting reports from 1995 to 2015, reports obtained from other international fishery organizations such as the Food and Agriculture Organization of the United Nations, and other peer-reviewed published papers, and will identify a

---

[1] Process tracing can contribute decisively both to describing political and social phenomena and to evaluating causal claims and it has been used as an essential form of within-case analysis.

© The Author(s), under exclusive license to Springer Nature
Switzerland AG 2021
L. R. Gonçalves, *Regional Fisheries Management Organizations*,
https://doi.org/10.1007/978-3-030-70362-2_4

possible formation of an epistemic community, and evaluate if in this case, they influenced the policy decisions on quotas for SBT management.

The CCSBT and "the Convention"

In the mid-1980s, it became apparent that the SBT stock was at a level where management and conservation were required. There was a need for a mechanism to limit catches. The main nations fishing SBT at the time, Australia, Japan, and New Zealand, began to apply strict quotas to their fishing fleets from 1985 as a management and conservation measure to enable the SBT stocks to rebuild.

On 20 May 1994, the then-existing voluntary management arrangement between Australia, Japan, and New Zealand was formalized when the Convention for the Conservation of Southern Bluefin Tuna, which had been signed by the three countries in May 1993, came into force. The Convention created the Commission for the Conservation of Southern Bluefin Tuna (CCSBT).

Other fishing nations were active in the SBT fishery, which reduced the effectiveness of the member's conservation and management measures. The principal non-member nations were Korea, Taiwan, and Indonesia. There were also a number of other fishing vessels flying flags of convenience, which operated in the fishery. As a matter of policy, the CCSBT has encouraged the membership of these countries.

The Republic of Korea and Indonesia joined the Commission on 17 October 2001 and 8 April 2008, respectively. The Fishing Entity of Taiwan's membership of the Extended Commission became effective on 30 August 2002.

At its meeting in October 2003, the CCSBT agreed to invite countries with an interest in the fishery to participate in its activities as formal cooperating non-members. Cooperating non-members participate fully in the business of the CCSBT but cannot vote. Acceptance as a cooperating non-member requires adherence to the management and conservation objectives of the CCSBT and agreed catch limits. Cooperating non-member status is regarded as a transitional measure to full membership and accession to the Convention.

The Philippines, South Africa, and the European Community were formally accepted as cooperating non-members on 2 August 2004, 24 August 2006, and 13 October 2006, respectively.

The object of the Convention is to ensure, through appropriate management, the conservation and optimum utilization of SBT.[2]

An interesting feature of the Convention and how it differs from most RFMOS and from those considered in this book, is that it does not have a geographical area—it applies to SBT in all oceans, including the spawning ground south of Java, Indonesia.

---

[2] The Convention—Article 3 http://www.ccsbt.org/userfiles/file/docs_english/basic_documents/convention.pdf.

Where the CCSBT mandate overlaps with other RFMOs, the CCSBT has had agreements or MOUs (Memorandum of Understanding) with these RFMOs, which clarify that the CCSBT has primary competence for the management of SBT.

The Convention established the CCSBT and describes how it operates and functions. The functions of the CCSBT include collecting information, deciding on a total allowable catch (TAC) and its allocation, deciding on additional measures, agreeing an annual budget, and encouraging accession by other states (as described in Article 8[40]).

Membership of the CCSBT is only open to States. To facilitate the participation of fishing entities, the CCSBT established the ECCSBT and the ESC in 2001. Fishing entities may be admitted as members of the ECCSBT and the ESC, and the fishing entity of Taiwan was so admitted in 2002. Membership of the ECCSBT and the ESC also includes all parties to the Convention.

The ECCSBT and the ESC perform the same functions as the CCSBT and the Scientific Committee (SC), respectively, with each member having equal voting rights. Decisions of the ECCSBT, which are reported to the CCSBT, become decisions of the CCSBT unless the CCSBT agrees otherwise.

Any decision of the Commission that affects the operation of the ECCSBT or the rights, obligation, or status of any individual member within the ECCSBT should not be taken without prior due deliberation of that issue by the ECCSBT.

Currently, the ECCSBT consists of six members and three cooperating non-members (Table 4.1).

The CCSBT has five subsidiary bodies which provide advice on their areas of expertise—the Scientific Committee (SC)/Extended Scientific Committee (ESC), Stock Assessment Group (SAG), Ecologically Related Species Working Group (ERSWG), Compliance Committee (CC), and the Finance and Administration Committee (FAC). A panel of independent scientists (the independent advisory panel) also sits in on the SC and SAG meetings and are able to provide advice directly to the CCSBT if required.

The diagram below shows the relationships between the CCSBT, its subsidiary bodies, and the Secretariat (Fig. 4.1).

The SC's main role is to assess and analyze the status and trends of the population of SBT and report and make recommendations to the CCSBT.

Two important working groups relate to the SC:

Table 4.1 List of CCSBT and ECCSBT members

ECCSBT members	Cooperating non-members
Japan	Philippines
Australia	South Africa
New Zealand	European Union
Republic of Korea	
Fishing entity of Taiwan (only ECCSBT)	
Indonesia	

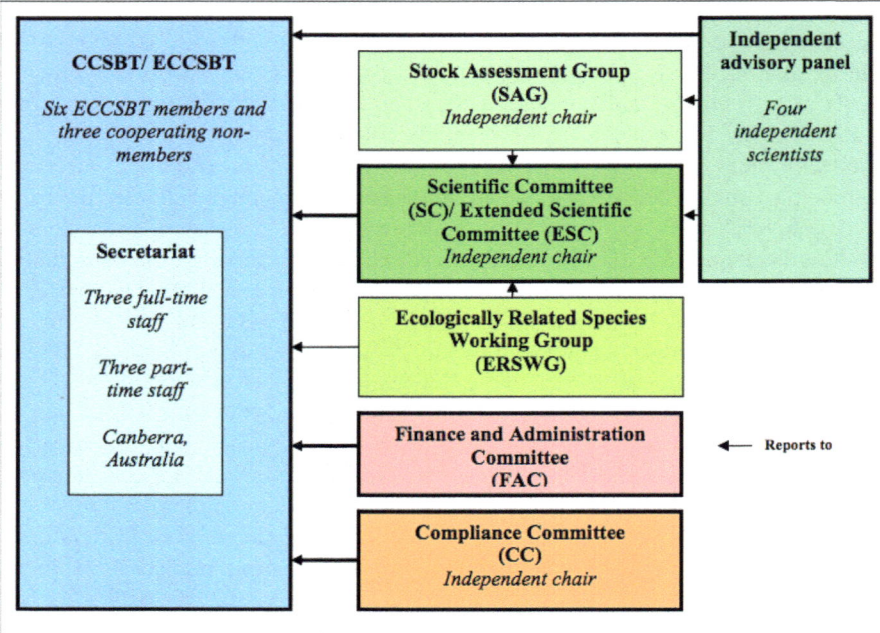

Fig. 4.1 The CCSBT organogram shows the relationships between the CCSBT, its subsidiary bodies, and the Secretariat

- the Stock Assessment Group (SAG) which was established to carry out technical evaluation functions including reviewing any new information on the SBT stock and updating the stock assessment.
- the Ecologically Related Species Working Group (ERSWG) which was established to provide information and advice on issues relating to species associated with SBT with specific reference to:

  (a) species (both fish and non-fish) which may be affected by SBT fisheries operations.
  (b) predator and prey species which may affect the condition of the SBT stock.

The Southern Bluefin Tuna

Southern Bluefin tuna (*Thunnus maccoyii*) (SBT) are large, fast swimming, pelagic fish (i.e., living in the open seas). SBT are found throughout the Southern Hemisphere mainly in waters between 30 and 50° south, but only rarely in the Eastern Pacific. The only known breeding area is in the Indian Ocean, south-east of Java, Indonesia.

SBT can live for up to 40 years, reach a weight of over 200 km, and measure more than 2 m in length. There is some uncertainty about the size and age when on average they become mature. This is the subject of current research by Commission members. The available data suggests that it is around 1.5 m and no younger than age 8. Mature females produce several million or more eggs in a single spawning period (CCSBT 2008).

Breeding takes place from September to April in warm waters south of Java. The juveniles migrate south down the west coast of Australia. During the summer months (December–April), they tend to congregate near the surface in the coastal waters off the southern coast of Australia and spend their winters in deeper, temperate oceanic waters. After age 5, they are seldom found in near shore surface waters (CCSBT 2008).

As SBT breed in the one area (south of Java) and all look alike wherever they are found, they are managed as one breeding stock.

Southern Bluefin tuna are very valuable and their primary market is the Japanese Sashimi market. Because of the high fat content of SBT flesh, premium prices can be obtained in the Japanese market. The total value of the SBT global fishery is estimated to be about $AUD1 billion (CCSBT 2008).

Except for the catch by Australian fishers, the main method used for catching SBT is longline fishing. This method involves using long lengths of fishing line with many hooks. The SBT caught are mainly frozen at very low temperatures ($-60°$ C) and either unloaded at intermediate ports and shipped to markets in Japan or unloaded directly at markets in Japan.

The Australian component of the fishery mainly uses the purse seine method. This is a net that encloses a school of fish. However, rather than landing the fish, the fish are towed to waters near the Australian mainland and placed in floating cages anchored to the ocean floor. The tuna are then fattened for several months and sold direct to Japanese markets as frozen or chilled fish (CCSBT 2008).

The Management and Conservation of SBT

Following entry into force of the Convention in 1994, the CCSBT was able to reach agreement on a TAC and quota allocations among its three original members (Australia, Japan, and New Zealand) the very next year (for the fishing year 1995–1996). The following year, the CCSBT also agreed on a TAC and quota allocations, as well as on a data collection and exchange program and certain other measures. However, the TAC reached did not necessarily reflect consensus. From those past years, Australia and New Zealand were much more conservative and precautionary on TAC than Japan (CCSBT 1996).

Beginning in 1997, however, trouble arose. According to the CCSBT (1997b) report, in the CCSBT Fourth Meeting (page 7, part 2) Japan proposed an increase in TAC of 3000 t, noting that this was in addition to any quota offer to non-members. Japan's proposal was based on the projections made by Japanese scientists of future stock size, which showed a high probability that the parental stock would recover to 1980 levels by 2020.

Australia stated that given the severely depleted stock status, it was vital that the precautionary approach be taken in relation to the TAC and national allocations. Australian scientists' assessments showed that the SBT stock was likely to remain in a depleted state for many years.

New Zealand stated its concern over the depleted stock status as evidenced by the consensus decision in the 1996 scientific report that the SBT stock was at 5–8% of the 1960 level of parental biomass and 25–38% of 1980 level of parental biomass. Both New Zealand and Australian scientists estimated that the probability was less than 30% that the stock would recover to the 1980 level of parental biomass by 2020.

Therefore, given the lack of any progress towards restraining the catch of non-members, New Zealand proposed to reduce the TAC by 3000 t. This would be a substantial step towards achieving the objective of stock recovery to the 1980 level of parental biomass by 2020. New Zealand noted Japan had suggested that New Zealand consider unilateral catch reductions. However, New Zealand firmly believed that it was the Commission's responsibility to address the serious status of the SBT stock. Unilateral catch reductions would disadvantage New Zealand's strategic position in the Commission: it would disadvantage NZ industry; and it would not have an appreciable impact on overall removals as New Zealand's proportion of the total SBT catch was less than 3% of total removals. New Zealand urged parties to seriously consider how catch restraint could be achieved at this meeting.

Japan responded that it could not accept New Zealand's proposal for a 3000 t reduction, as it believed the stock could sustain a 3000 t increase. It was the view of Japan that taking an extreme precautionary approach, such as stopping fishing, did not reflect Japan's interpretation of stock status. As there were scientific uncertainties regarding SBT stock status, in Japan's view, it was necessary to work on research programs that would resolve one of the major uncertainties. New Zealand responded by pointing out that the precautionary approach, as expressed in the UNFSA, was not ambiguous—the absence of scientific data did not provide a rationale for failing to take conservation and management measures. In New Zealand's view, Japan was suggesting that uncertainty in the stock assessment meant that a catch increase was an acceptable action. However, a catch increase was clearly inconsistent with the precautionary approach expressed in UNIA and was irresponsible, given the current stock status.

Australia noted that the precautionary approach was clearly defined in both UNFSA and the FAO Code of Conduct on Responsible Fishing and was widely used in fisheries, and thus directly relevant to matters before the Commission.

The dilemma between Japan and Australia and New Zealand caused frustration in the countries, and they left the 1997 meeting failing to adopt a national TAC. The matter continued for the next several years.

The CCSBT did not agree on a global TAC or quota allocations for the fishing years 1997–1998, 1998–1999, 1999–2000, 2000–2001, 2001–2002, or 2002–2003. During this period, the Commission did make limited progress in certain other areas, including the creation of an independent panel of scientists and independent chairs for its scientific committees and the adoption of a Trade Information Scheme (TIS) in 2000.

Throughout those years, the CCSBT Scientific Committee would improve the information it shared; however, the level of uncertainty was still great. Scientists were working to quickly develop a management procedure that would, at the very least, guarantee a major certainty for the management advice, as uncertainty was the major excuse for not accepting the scientific precautionary advice.

The Scientific Committee and the Management Procedure

During the 1990s, managers from Australia, Japan, and New Zealand consistently had diverging opinions about the interpretation of the scientific advice and a disagreement about the setting of the TAC (e.g., CCSBT 1995a, b, 1996, 1997a, b, 1998, 1999a, b). Initially, the stock assessments tended to predict a high probability of rapid stock rebuilding. However, as the years passed, the projected rebuilding was not evident from the data and different assessments yielded divergent predictions (e.g., Klaer et al. 1996). This led to a number of initiatives to attempt to improve the stock assessment, including the development of a joint experimental fishing program (Polacheck 2002 and references therein). When agreement on this failed, Japan undertook unilateral experimental fishing, which was viewed by other CCSBT members primarily as a means of increasing catches. This resulted in a legal dispute in the International Tribunal for the Law of the Sea (ITLOS) (Firestone and Polacheck 2003). The ITLOS hearing resulted in a temporary suspension of Japanese experimental fishing in 2000; however, the Arbitral Panel subsequently ruled that it did not have legal jurisdiction to resolve the case.

During the 1990s, catches of SBT by non-CCSBT parties (principally Taiwan, Korea, and Indonesia) increased to substantial levels, adding impetus for the Commission to resolve its problems.

With the Convention's entry into force in 1994, the SC was established and the tripartite science process that had existed between Australia, Japan, and New Zealand was formalized. The SC first met in 1995 and has met annually since then.

The first working group of the SC, the ERSWG, was established early on in the history of the CCSBT. It had its first meeting in 1995. The SAG followed in 1998 when the CCSBT separated the observing and analyzing functions from the interpreting and advising functions. The SAG reports to the SC and the ERSWG terms of reference provide that it also reports to the SC, but in practice, it reports directly to the CCSBT.

In 2000, the CCSBT members negotiated a settlement for the experimental fishing dispute. It also agreed to: (1) the appointment of an independent scientific advisory panel; (2) development of a Scientific Research Program aimed at improving data for stock assessments; and (3) the development of a management procedure. The concept of a simulation-tested, management decision rule had been formally introduced into the CCSBT process in 1993 (Sainsbury and Polacheck 1993) and the CCSBT agreed to hold a Management Strategy Workshop in 1996 (CCSBT 1996). However, the first CCSBT Management Strategy Workshop was not held until 1999 (CCSBT 2000). Prior to this, an in-depth illustration of the applicability of an MP approach to the SBT stock had been completed (e.g., Polacheck et al. 1999). The Commission agreed to develop an MP in 2000 (CCSBT 2000).

In 2001, the Extended Scientific Committee (ESC) was established to provide for participation by the fishing entity of Taiwan and other cooperating non-members.

An important feature of the CCSBT science process is the role played by both the independent chairs of the SAG and the SC, and the advisory independent panel. These roles were established following recommendations of a group of independent stock assessment and scientific fishery advisers who were asked by the CCSBT to evaluate its science processes and methods.

The independent chairs of the SAG and the SC draft meeting agendas, direct discussions to ensure good scientific principles are observed, facilitate consensus, and carry out other activities as chair of the meetings.

The advisory panel participates in all meetings of the SAG, SC, and other scientific meetings. Their role is to help consolidate parties' views to facilitate consensus and their views are incorporated in SAG/SC reports. They also provide their own views on stock assessments to the SC and CCSBT.

The situation seems to have improved in 2003. By that year, the Republic of Korea had joined the CCSBT and there was also created an "extended" Commission to facilitate the participation of Taiwan. In 2003 and 2004, the CCSBT reached agreement on TACs and quota allocations for its members and certain non-members, and made progress in starting to develop a comprehensive "management procedure" that the Commission ultimately adopted in 2011 (see below).

In 2005, however, the CCSBT received advice from its Scientific Committee that the stock of SBT was deteriorating, that at current catch levels there was a 50% chance that stock levels would decline to zero, and that only very significant reductions in the TAC would result in a 50% probability of avoiding further decline. Despite this daunting advice, the CCSBT could not reach agreement on a TAC nor any reductions in catch levels. Instead, members and cooperating non-members merely promised that their individual catch levels would not exceed those from the previous year.

In 2006, the CCSBT faced a true crisis. As described by the CCSBT (2008), the CCSBT "considered information that catches over the past 10–20 years may have been substantially underreported and the implications that had for the historical data record maintained by the CCSBT." This revelation of substantial underreporting of catches may have seriously compromised the data on which the Commission must make decisions even now and well into the future.

At the same year, the CCSBT considered information that catches over the past 10–20 years may have been substantially underreported and the implications that had for the historical data record maintained by the CCSBT. These uncertainties in the historical catch and Catch per Unit Effort (CPUE) for SBT have made it difficult to run a full stock assessment.

In 2006, the CCSBT adopted a TAC for most members for 2007–2009 (2007–2011 in the case of Japan) that was only to be reviewed if exceptional circumstances emerged in relation to the stock. During the three-year fixed-TAC period, the ESC and the SAG are focusing on reducing uncertainty in the data upon which the SBT stock assessment is based with the intention of conducting a full stock assessment in 2009.

According to the CCSBT (2008) "The estimates of the depletion of the spawning stock biomass suggest that, in terms of outcomes, the CCSBT has not been successful in managing SBT. In addition, due to the uncertainty in past underreported catches, the data holdings of the CCSBT are compromised and their utility for scientific stock assessment to inform management decisions is significantly diminished. Nonetheless, the ESC, including an independent advisory panel, has sought to provide the Extended Commission with the best scientific advice possible on the status of the SBT."

Perhaps chastened by this crisis, the CCSBT agreed in 2006 to a global TAC and national allocations that included a reduction in the Japanese allocation from 6065 t to 3000 t (quota set for 5 years until at least 2011, pending a review). The following year, the CCSBT set a global TAC and allocations for the 2008–2009 period.

The CCSBT is struggling to reduce the uncertainty in the data on which the SBT stock assessment is based. At present, though, the CCSBT still faces glaring problems arising from the compromised data, including the admission in the CCSBT (2008) that "it is not possible to determine the exact trends in the status of SBT over time."

According to Kolody et al. (2008), one of the greatest benefits of the CCSBT MP process was the facilitation of these stakeholder interactions. It helped to increase the general understanding of the fishery for all involved, and provided the mechanism and context for feedback between stakeholders and MP developers. It also increased the shared knowledge between stakeholders and decision-makers. It would be unfortunate if the recent revelations of the historical data problems have undermined the benefits of communication and cooperation that were gained through these consultations.

For Kurota et al. (2010), another strong point of the MPs is to promote transparent and rapid decision-making for TACs. In principle, if an MP becomes available, TACs are determined automatically, so that confrontation among scientists, managers, and stakeholders can be avoided; transparency in the process promotes agreement on management measures by removing skepticism amongst stakeholders. In addition, rapid decision-making with a short time lag between data collection and implementation is critical to secure sustainable use of marine resources.

At its eighteenth annual meeting (2011), the CCSBT agreed that a Management Procedure (MP), known as the "Bali Procedure," would be used to guide the setting of the SBT global TAC to ensure that the SBT spawning stock biomass achieves the interim rebuilding target of 20% of the original spawning stock biomass. The CCSBT now sets the TAC based on the outcome of the MP, unless the CCSBT decides otherwise based on information that is not otherwise incorporated into the MP. This was a great step forward.

In adopting the MP, the CCSBT emphasized the need to take a precautionary approach to increase the likelihood of the spawning stock rebuilding in the short term and to provide industry with more stability in the TAC (i.e., to reduce the probability of future TAC decreases).

An MP is a pre-agreed set of rules that can specify changes to the TAC based on updated monitoring data. An MP is defined as a simulation-tested decision rule (or Harvest Control Rule), and the requisite methods of data collection and analysis, which together are used to calculate a management recommendation (e.g., Total Allowable Catch (TAC)) for a fishery (e.g., de la Mare 1986; Butterworth et al. 1996; Smith et al. 1999).

MP development was carried out by CCSBT SC members, who consist of about 10–20 "national" scientists in member countries, an independent advisory panel (4 experienced specialists), and a consultant programmer. The advisory panel for the SC was officially established in 1999 to facilitate consensus among member countries. While a set of operating models (OMs) was developed jointly by the SC, candidate MPs were proposed by scientists of member countries (Kurota et al. 2010).

From 2002 to 2011, the CCSBT conducted extensive work to develop an MP in order to guide its global TAC setting process for southern Bluefin tuna. The CCSBT tested a variety of candidate MPs with the aid of an operating model of the fishery that simulated the characteristics of the SBT stock and fishery. The candidate MPs were tested against a range of uncertainties so that a robust procedure could be identified.

Considerable optimism existed within the CCSBT that a jointly developed MP could break the dysfunctional cycle of contested stock assessments and failure to reach consensus on management decisions that had prevailed since the mid-1990s. However, the development process suffered from a number of setbacks, culminating in revelations of substantial data problems in 2005–2006 that undermined confidence in the agreed MP. As a result, MP implementation was suspended until the implications of the data problems can be formally admitted within the simulation testing process.

The final MP, known as the "Bali Procedure," was recommended by the CCSBT's Scientific Committee in July 2011. Parameters of the recommended decision rule could be adjusted to set different time horizons for rebuilding, and to constrain the maximum TAC changes allowed each time the TAC is updated. A range of options was presented to CCSBT's Extended Commission.

The Extended Commission adopted the Bali Procedure and the following associated management parameters as its MP at the CCSBT's eighteenth annual meeting in October 2011:

• The MP is tuned to a 70% probability of rebuilding the stock to the interim rebuilding target reference point of 20% of the original spawning stock biomass by 2035.
• The minimum TAC change (increase or decrease) is 100 t.
• The maximum TAC change (increase or decrease) is 3000 t.

- The TAC will be set for three-year periods, subject to paragraph 7 of the Resolution on Adoption of a Management Procedure.[3]
- The national allocation of the TAC within each three-year period will be apportioned according to the Resolution on the Allocation of the Global Total Allowable Catch.[4]

The MP has been used to guide the setting of the global SBT TAC for fishing years since 2012. For the second (2015–2017) and subsequent three-year TAC setting periods, there is a one-year lag between TAC calculations by the MP and implementation of that TAC (e.g., the 2015–2017 TAC was calculated in 2013).

The technical specifications of the MP may be updated, and the last review was, to date, in 2013.[5]

According to the CCSBT (2014) at the Independent Review, the advice presently delivered to CCSBT by its scientific subsidiary bodies is excellent. The institutional setup, with its independent panels and chairs, the systematic peer-review processes, the adoption of instruments like the MP, the metarule, triennial in-depth assessments, indicators, etc., provide instruments which are at the top of international standards. The advice delivered by the ESC has apparently always been followed up.

The Epistemic Community at CCSBT

The CCSBT, besides being the newest RFMO in this study, has a very well structured scientific panel, which was responsible for building a management procedure that was coherent; it was subsequently accepted by the ECCCSBT.

Before 2011, Japan, Australia, and New Zealand had zero, or almost null confidence, in scientific information. They were feeling that the stocks were declining, but they did not know how much, or how severe, or when, or even how much they had left. In every meeting they had, they were clamoring for more certainty. As Clark and Majone (1985) explained, and as demonstrated empirically in the Haas and Stevens (2011) chapter: when science is not fully safeguarded from politics, in spite of the quality of the science produced, it may not be considered true by many of the decision-makers.

Since the establishment of the ESC, SAG and the independent advisory panel, the information began to grow clearer and the need for reducing the quotas, for the time being, and in the near future have a better quota system in place, was accepted.

In 1997, when a crisis occurred within the CCSBT, it came to light that the data available on SBT and the techniques used by scientists to analyze this data were

---

[3] http://www.ccsbt.org/userfiles/file/docs_english/operational_resolutions/Resolution_Management_Procedure.pdf.

[4] http://www.ccsbt.org/userfiles/file/docs_english/operational_resolutions/Resolution_Allocation.pdf.

[5] http://www.ccsbt.org/userfiles/file/docs_english/general/MP_Specifications.pdf.

comparable to those utilized elsewhere in the world, although scientists' representatives had some concerns about the data and models used for assessing SBT. Their major concern was with the process and group dynamics that lead to the report provided to the Commission for the Conservation of Southern Bluefin Tuna (CSBT) and the lack of agreement on what advice should be provided.

The process observed in the Stock Assessment Group (SAG) and the Scientific Committee (SC) could not be described as scientifically neutral. In other independent scientific processes, scientists participate firstly as individuals, not as national representatives.

In this context, a recommendation arose showing the necessity of a clear separation between science and management. Separating the process into a technical part (the SAG), and an advisory part, (the SC), may be a first step in that direction, but would not appear to be sufficient under present conditions. As an interim measure, the CCSBT constituted a facilitating panel of three to five independent scientists to guide the SAG/SC process towards consensus advice. If the SAG/SC could not reach consensus, the panel itself would provide the advice to the CCSBT.

This process approved in 1999[6] was a major step forward. The scientists to be selected for the independent chairs of the SAG and SC, as well as for the members of the Advisory Panel (CCSBT, 1999–2001, part I):

- Should not be a national of the parties nor have been a permanent resident or have worked for the parties since 31/12/89 except where parties reach a consensus to choose the qualified individual.
- Must have an excellent technical ability in stock assessment.
- Must have adequate working experience as a scientist involved in stock assessment and fisheries management at the international level.
- Should have working experience with large pelagic fish resources.
- Are desired to have familiarity with assessment procedures and scientific procedures used in international fishery commissions.

Also the duties of the independent chair of the SC are to[7]:

- circulate the draft agenda through the Secretariat.
- declare the meeting schedule, and open and close meetings of the SC.
- direct discussions in the meetings, to ensure that the work of the SC adheres to the scientific principles of demonstrable evidence, statement of assumptions and examination of logic, and to ensure observance of these rules.
- facilitate reaching consensus to the extent possible.
- accord the right to speak and to limit the time allowed for speaking.
- rule on points of order, subject to the right of any member to request that any ruling by the Chair be submitted to the SC for decision.
- ascertain if consensus exists.

---

[6] http://www.ccsbt.org/userfiles/file/docs_english/operational_resolutions/report_of_the_1998_peer_review_panel.pdf.

[7] http://www.ccsbt.org/site/stock_assessment.php.

- in relation to each meeting of the SC, to sign, on behalf of the SC, a report of the proceedings of the meeting for transmission to the Commission and present such reports to the Commission.
- to convey to the Executive Secretary any instructions determined by the SC.
- the authority to call meetings of the representatives of the members after conferring with the representatives.
- to exercise other powers and responsibilities as provided in the Rules of Procedure for the Commission and make such decisions and give such directions as will ensure that the business of the SC is carried out effectively and in accordance with its decisions.

The Advisory Panel terms of reference are:

- to participate in all meetings of the SC and other scientific meetings as requested by the Commission.
- to help to consolidate parties' views to facilitate consensus.
- to incorporate their views in SC reports and provide to the SC and CCSBT in the form of a report of their own views on stock assessment and other matters.

All the measures were taken to guarantee a more independent view of the process, and to give more confidence between countries members, as the confidence between them was never that strong, as proved by the legal dispute undertaken by Australia and New Zealand against Japan (1998) under the International Tribunal for the Law of the Sea (ITLOS), mentioned above.

In this new scientific structure, some essential characteristics were guaranteed, which very much aligns them with what Haas and Stevens (2011) considered important to promote policy changes. The more autonomous and independent science is from policy, the greater its potential influence (Andresen 2000; Botcheva 2001; Haas 2001). Consensus in isolation builds value and integrity, its consequences, then, should be discussed publicly. Measures of autonomy and integrity include the selection and funding of scientists by intergovernmental organizations rather than by governments, the recruitment of scientists by merit on important panels, and reliance on individuals whose reputation and authority rest on their role as active researchers rather than on their role as policy advocates or science administrators. Accuracy can be achieved via peer review, interdisciplinary research, and independence from sponsoring sources.

And, the promotion of what is called usable knowledge[8] usually happens through the "epistemic communities." In the CCSBT case, this specific community had not emerged until this new design as the scientists could not reach consensus nor any agreement on methods, results, and even more in their policy recommendations. In

---

[8]Already defined in the previous chapter, but to facilitate the reading "In short, usable knowledge encompasses a substantive core that makes it usable for policymakers and a procedural dimension that provides a mechanism for its transmission from the scientific community to the policy world, and it allows for agency when theorizing about broader patterns of social learning, policymaking, and international relations (Haas and Stevens 2011)."

addition, before the implementation of independent chairs and an Advisory Panel, there was no confidence in the research results provided by the countries themselves.

However, even after the implementation of independent chairs and an Advisory Panel, there was still no agreement on quotas. The level of confidence and agreement increased, but there was no influence on political decisions for quotas. The agreement existing so far came only after the establishment of the MP (CCSBT 2011). When the implementation of the MP happened, the rule was clear, a scientific baseline was defined, and from then on, all political decisions have been made based on this rule.

When the MP was approved, the CCSBT was able to overcome one of the most common problems for the RFMO's effectiveness, which has already been pointed out by Gjerde et al. (2013). According to them, countries are normally interested in enhancing or maintaining their domestic fishing opportunities, leading to the pursuit of short-term gains over long-term sustainable fishing. And, this was the exact situation at CCSBT, before the MP, when it was mainly Japan who was fighting to increase quotas. Australia and New Zealand, who wanted to keep fishing, were looking towards the horizon in their attempts to establish more conservative quotas.

The epistemic community within the Stock Assessment Group, the Extended Scientific Group, and the Independent Advisory panel formed a trusted body, which countries could rely on. But more than that, as it is an established and approved procedure, the decision-makers no longer doubt their decisions, they need only to follow the rules.

In this case, the CCSBT differs from the previous RFMOs studied here: since the beginning, in the CCAMLR, knowledge flowed through the established institutional design, while in the case of the ICCAT there were no clear procedures and the epistemic community only had an effect when multiple stakeholders acted together. The CCSBT epistemic community is a mixture of both of these processes. They first acted together to create a reliable mechanism that would be accepted by all countries, and then, they had a clear institutional design where usable knowledge can feed directly into the political decisions.

NGOs at CCSBT

The CCSBT process for acceptance of observer status for NGOs was never too clear and/or too open and transparent.

Since 1997, for instance, Greenpeace International requested access to the meetings and was denied. The Fourth Annual Meeting of the Commission (CCSBT 1997b, p. 2) in Japan stated that "it did not support the application because Greenpeace was known to use radical methods to achieve its goals and to act against the provisions and spirit of the Convention. Japan stated that in light of the Commission's objective to ensure, through appropriate management, the conservation and optimum utilization of southern bluefin tuna, it would be inappropriate to assign observer status to Greenpeace."

Even with the complaints from New Zealand and Australia, who promptly disagreed with Japan noting that the Commission's proceedings should also be transparent and if there were good reason to facilitate observation of the meeting by NGOs with a legitimate interest in the management of SBT, the observer status for Greenpeace was denied from 1997 to 2000. The strategy Greenpeace used to follow the meetings was to be part of a country's delegation. Thus, Greenpeace had members as part of the Australian and New Zealand delegation in 2000 and 2001. And, in 2001, they also had a member as part of the Australian Scientific Delegation. Of course, when an NGO is a member of a country's delegation, they are wearing more than a government hat, more, in fact, than an NGO hat; however, this is still very helpful because they can lend their expertise to the government while also having access to inside information.

In 1997 during Japan's opening statement, the question was again brought up about environmental groups: "as we can see from the recently published report by WWF and Traffic Oceania, as well as the application for the observer status from Greenpeace International, the general public's interest in CCSBT's activities is increasing. Claims made by these groups are based solely on pessimistic information. We are concerned that not only do they fail to reflect CCSBT's views with fairness, but run the risk of creating misunderstanding and confusion within the society by stirring up excessive sense of crisis. Faced with this situation, Japan believes that CCSBT needs to demonstrate that we are a regional fisheries management organization, which is functioning effectively. In this context, it is critical for us to swiftly achieve substantive progress on major issues I talked about earlier, which will, in turn, increase the credibility of our forum" (CCSBT 1997b).

Based on Japan's statement, it is important to note that pressure from environmental groups was influencing the decision-makers to improve the effectiveness of CCSBT, even while operating from the outside.

Besides the Japan resistance against Greenpeace International, other NGOs, such as TRAFFIC,[9] were following the meetings either as scientific experts or as part of the Australian delegation since 1996, including the Tuna Boat Owners Association of Australia.[10]

However, it was only in 2007 that the CCSBT finally accepted NGOs as observers. The main difference in this case is that as observers, the organizations now had a forum to make their own statements.

The Human Society International, WWF, and TRAFFIC are, since 2007, following the meetings, and with this the CCSBT no doubt has guaranteed more transparency to the entire process. In 2012, after the approval of the MP, TRAFFIC and WWF welcomed the decision.

---

[9] TRAFFIC aims to ensure that trade in wild plants and animals is not a threat to the conservation of nature (http://www.traffic.org/).

[10] ASBTIA is the peak body representing Southern Bluefin Tuna ranching companies in Australia. The Association has 13 member companies, which is 100% of the local industry. The Association began in 1978 as the Tuna Boat Owners of South Australia (http://asbtia.com.au/about/).

Although the NGOs are now allowed to be part of the CCSBT Annual meetings, it is possible to note that their voices are still not as vocal as are the other NGOs in the CCAMLR and ICCAT. There are many reasons that may explain that, such as the difficulty for international organizations to follow up on meetings that normally happen in the Oceania area, far from their headquarters, or possibly even due to the lack of resources or prioritization to be involved in so many international agreements.

In the CCSBT case study, the importance of the Scientific Committee is even greater, as their independent view guarantees a more insulated process and more reliable data, as civil society has not been successfully able to follow the matters of SBT.

Conclusion

The CCSBT is the youngest and smallest of the RFMOs studied here. However, it has made huge improvements in its scientific process, which has led to a robust procedure in supporting countries on their decisions.

Since its governance crisis in 1997–1999, the CCSBT has made progress, some of which is already evident in the 2008 performance review assessment (CCSBT 2008), improving its institutional structure and functioning, its information systems and assessment methods, its decision-process and compliance monitoring systems.

However, the approval of the MP in 2011 was a great turning point and proved that CCSBT countries were really up to change *modus operandi* to promote a better SBT management. It was a shift from countries taking over the decisions, to now, countries accepting these decisions based on science that has been provided by a well-structured and independent epistemic community. It would be interesting to see in the near future a review of Cullis-Suzuki and Pauly's (2010) quantitative research, with data post-MP, to evaluate if they would get a higher score, as the CCSBT earned the lowest score amongst all RFMOs.

The epistemic community emerged only in 2011/2012, when the MP was adopted and the Independent Advisory Panel, the SAG and the ESC, which constitute the epistemic community, were acting on behalf of an insulated science giving trustworthy subsidies to the ECCSBT.

At its Eighteenth annual meeting in 2011, the CCSBT agreed that an MP would be used to guide the setting of the SBT global total allowable catch (TAC) to ensure that the SBT spawning stock biomass achieves the interim rebuilding target of 20% of the original spawning stock biomass. In adopting the MP, the CCSBT emphasized the need to take a precautionary approach to increase the likelihood of the spawning stock rebuilding in the short term and to provide industry with more stability in the TAC (i.e., to reduce the probability of future TAC decreases).

It is still too early to see results from the MP on the stock status. Thus, the stock remains at a very low state estimated to be 9% of the initial SSB, and below the level to produce maximum sustainable yields (MSY); however, there has been some improvement since the 2011 stock assessment and fishing mortality is below the level associated with MSY—considered already progress.

Also, on a positive note, the membership of the CCSBT has grown in recent years such that the members and cooperating non-members of the Commission now represent virtually all-fishing activity for SBT, conferring more governance and regulation for the system. With all the relevant players now at the table, the CCSBT at least has an opportunity to create a better future for SBT and those whose livelihoods depend on that resource.

CCSBT is still a new RFMO, but it has so far demonstrated a strong institutional design where science can influence political decisions, and the epistemic community can provide the usable knowledge coherent enough to guide political decisions, even considering a level of uncertainty.

CCSBT has yet to demonstrate more commitment with transparency and with the isolation of policy to science. However, for a new RFMO with a fairly small membership whose original mandate is limited to ensuring the conservation and optimum utilization of a single fish stock, this is already a big step.

Conclusion

A fundamental principle underlying successful modern fisheries management is that management decisions need to be based on the "best available scientific information," and carried out through cooperation between different stakeholders.

The United Nations Convention on the Law of the Sea (UNCLOS) highlights this principle stating that the determination of allowable catches and other conservation measures for living resources in the high seas are to be based on the "best scientific information available to the States concerned"[1] and require states to "cooperate to establish subregional or regional fisheries organizations to this end" (RFMOs).[2]

The need for scientific advice as the basis for management decisions and the establishment of RFMOs is further affirmed in the United Nation Fish Stock Agreement (UNFSA).[3]UNFSA also includes provisions with respect to transparency and dissemination of information. In this respect, the agreement mandates that states and RFMOs "shall actively promote the publication and dissemination to any interested States the results of that research and information relating to its objectives and methods".[4] It also requires that "States shall provide for transparency in the decision-making process and other activities of subregional and regional fisheries management organizations and arrangements".[5]

Even with all of these rules and norms, the main question posed here from the very beginning still remains: is power listening to science on RFMOs?

And the answer is: they may listen to them, but the strategy, their allies, and the way it happens differ case by case (Table A.1). Also, it is important to note that, in

---

[1] Article 119—http://www.un.org/depts/los/convention_agreements/texts/unclos/part7.htm.

[2] *Op cit*—Article 118.

[3] http://www.un.org/Depts/los/convention_agreements/texts/fish_stocks_agreement/CONF164_37.htm.

[4] *Op cit*—Article 14.

[5] *Op cit*—Article 12.

L. R. Gonçalves, *Regional Fisheries Management Organizations*,
https://doi.org/10.1007/978-3-030-70362-2

Table A.1 A summary review of the case of studies

	Case CCAMLR	Case ICCAT	Case CCSBT
Signed on	1982	1966	1994
Cullis-Suzuki and Pauly (2010) ranking evaluation	100%	37.5%	0%
Number of members	$N = 25$ members +11 acceding states	$N = 80$	$N = 6 + 3$ cooperating non-members
Epistemic Community	Scientists ($n = 25$) from the Working Group Fish Stock Assessment (WG-FSA)	Some NGO members, ICCAT chairman, some SCRS members	Independent Advisory Panel, the SAG and the ESC
Main NGOs	ASOC, COLTO	Pew, Greenpeace, WWF	Traffic, Greenpeace
Strategy and allies	WG-FSA is formed by a high caliber of scientists, and based on their knowledge and sound science, they provide advice to the SC and their advice is automatically accepted by the Commission. The institutional design also supports science influence	The epicom had more influence when it used a crisis momentum and was backed up by the NGOs, CITES, and public opinion. In this case study, the ICCAT chairman was also part of the epicom and had access to the major players. Multiple stakeholders acted together. The epicom here acted, not through its own institutional design, but through its personal communication channels	CCSBT is a very new RFMO, and still not much open to NGOs participation, which would increase their transparency. In this case, the SC and the Independent Advisory Panel are important as they produce accurate knowledge to subsidize policy decisions. Their strategy was to approve a management procedure, when the implementation of MP happened, a rule was clear, a science baseline was defined, and from then on, all of the political decisions have been made based on this
Policy effect	TAC based on science + CDS adoption	TAC based on science	Management procedure— more autonomous and independent science and TAC based on science

(continued)

Table A.1 (continued)

	Case CCAMLR	Case ICCAT	Case CCSBT
Effectiveness outcome on biomass[a]	See Fig. 2.3. Possible stability around that level (50% of the biomass virgin) for the coming years	See Fig. 3.4	It is still too early to see results from the MP on the stock status. Thus, the stock remains at a very low state estimated to be 9% of the initial SSB, and below the level to produce maximum sustainable yield (MSY); however, there has been some improvement since the 2011 stock assessment and fishing mortality is below the level associated with MSY—considered already a progress

[a]Through this book and methodology it is not possible to state that a policy outcome, influenced by the epicom, had a direct effect on the biomass, as it is not possible to control other variables, such as natural variation and IUU control. However, their influence led to policies which credibly improved biomass. As illustrated in Chap. 1, the RFMOs used as case studies here have some form of a Scientific Committee that provides relevant scientific advice to the governing body on stock status, monitoring, and possible management actions (most frequently the setting of catch limits). The working arrangements of these Scientific Committees and their subsidiary technical bodies vary, for example, to the extent to which the Scientific Committees and their technical bodies undertake reviews of work prior to the meeting or actually undertake analyses during their meetings. And based on how they are structured, constituted and implemented, they differ on how influential they are on political decisions

a strict methodological sense, the facts cannot be generalized to other RFMOs on the basis of what is but a fairly modest sample, compared to the vast variety of the regional fisheries agreements that exists all across the oceans.

As shown, the CCAMLR presents the most established collective factors to guarantee that the knowledge can flow from its own epistemic community to the decision arena. As the agreement was historically built to follow science, there is no plea to not accept management advice. The epistemic community located within the WG-FSA provides data and advice to the Scientific Committee, and it arrives and it is accepted with no change at the annual meetings. This is a clear example of how the epistemic community can act through an agreement's institutional design and does not need any advocacy pressure to make it valid. It is not surprising that in the Cullis-Suzuki and Pauly (2010) paper, the CCAMLR scored better than other RFMOs. How they are implementing conservation measures, based on usable knowledge, guarantees that the best management advice was followed, and thus, the stocks maintain a good record. CCAMLR brings together the factors that Haas (2000) indicates as necessary for an effective agreement, such as "the existence and willingness of key states to exercise leadership, the existence of strong international institutions, the presence of a strong transnational scientific involvement (epistemic communities), and the involvement of NGOs."

The CCAMLR, for instance, presents many characteristics that differ from other RFMOs studied here and this more than justifies the opinion of some stakeholders being against considering CCAMLR as an RFMO; they prefer to call CCAMLR a "conservation body." The institutional design and the rules of procedure, where knowledge and social learning are considered on a higher level, are giving truth to power. The epistemic community at the CCAMLR has no difficulty in being listened to and after 10 years of documental analysis, their management advice was accepted *ipsis litteris*. They are trustworthy, they use the same scientific methods, and they act on behalf of knowledge where no rational strategy is needed to make their advice accepted. The empirical data evaluated here built the case around the Patagonian toothfish; however, I would hardly doubt that for other species it would be much different, as the rules are now followed by countries at every meeting.

ICCAT, in turn, does not have the same history, and on the contrary, it was largely known by a Commission that does not normally acknowledge science. It had, for a few years, a very bad reputation on managing the EBFT—one of the main and most important and emblematic species managed by them. However, 2009/2010 it was a great turning point. Although ICCAT's own institutional design was not aimed towards creating an insulated science, and with huge influence from politics on the species management clearly illustrated in Chap. 1, the EBFT case analyzed here showed that in a specific situation where there was a fishery crisis, a high level of uncertainty, and added public pressure made by a transnational network of NGOs and CITES, all helped to enhance the need for political will on the decision-makers. By this time, the epistemic community was ready to provide usable knowledge to the CPCs and in this case, it worked out well. After many years of not listening to science, the ICCAT CPCs accepted the SC's advice and now it is already possible to see signs of recovery of the EBFT populations. However, in order to make it happen it was necessary for lots of action from diverse stakeholders, it was necessary for an advocacy strategy behind the scenes—that which is not necessary at the CCAMLR, for instance. In ICCAT, the EBFT epistemic community is not limited to within the SC; it is formed by diverse stakeholders who are acting on behalf of knowledge, not through its own institutional design, but through their personal communication channels.

As the institutional design built for the ICCAT was not well designed to listen to science, as it was for CCAMLR, the EBFT, epistemic community was formed by a group with a high level of professionalism. It emerged from the SCRS and included scientists from NGOs, CPCs, and the ICCAT chairman, a scientist himself, producing usable knowledge, coherent, legitimate and solid enough to influence strategically, on many dimensions of power. The knowledge produced about the state of the EBFT population was sound enough to have other key stakeholders supporting the epistemic community claim. This major influence only worked due to the fact that NGOs and CITES widely publicized the emergent fishery crisis that could damage ICCAT's history. Moreover, they created pressure on ICCAT, sponsoring the epistemic community knowledge.

ICCAT does not work like the CCAMLR, however, the EBFT case analyzed in this book showed that when power does listen to science and implements their advice through a social learning process, fish stocks can recover, and thus reach the main goal of the Commission, conferring more effectiveness for the international agreement. If they can use EBFT as a lesson learned for the other stocks, it will be a huge step forward.

And last, but not least, while the CCSBT also presented an epistemic community, they only emerged when the Commission faced reality and acted to promote a new structure for science where independence from countries' opinions was guaranteed. When the Commission approved the Independent Chair and the Independent Advisory Panel, they were able to create the MP that was ultimately accepted by all countries and is now the ruling decisions on quotas. In the CCSBT case, the epistemic community resides within the SAG, ESC, and the Independent Advisory Group; however, they could only emerge when they gained confidence from countries showing that with the best science they had in hand, they were reducing the uncertainty, they were providing the best advice they could, under the management procedures they had created. It would be great to repeat the Cullis-Suzuki and Pauly (2010) research paper in 2020 (10 years later), to see if CCSBT's score would change now with the MP implemented and working.

Irrespective, in most cases, participation in the scientific process revolves around member delegations with official heads of delegations. The heads of delegation are most commonly government representatives with formal scientific credentials. The levels of relevant research experience and expertise in the fisheries being assessed, however, vary from almost nil to extensive. The head of a delegation has the formal responsibility for managing the inputs of the members of his or her delegation and for the delegation's agreement to the content of final reports of the scientific meetings. The functioning (e.g., procedures for participating in plenary and working group discussions) and structure of the national delegations varies substantially. Nevertheless, in most cases a large fraction of a member's national delegation is drawn from government fisheries departments or associated marine research institutes with a substantial portion of the funding for participation in the meetings and related research provided by the national fisheries management department or agency. It would be much better if they could retain an independent scientist as well.

The independence and separation of the scientific processes that provide information for consideration by the RFMO, through the policy deliberations of the RFMO, are important component of affecting the intent of UNCLOS and UNFSA whose management decisions are based on the best available scientific information.

The agreements history, the countries that are a part of it, the external context, as well as influence, the way knowledge is transmitted and structured, and the presence of a strong epistemic community made a lot of difference in the three cases presented here.

After observing the three cases, it is possible to draw some important conclusions that corroborate with Peter Haas' work from 1992, which was compiled and updated in Haas (2015).

Haas argues that consensus-based science can play an independent and important role by influencing and even reformulating state interests, thereby helping to bring about international agreements that transcend and reshape state interests. This is made possible through the involvement of experts. Thus, from the epistemic community perspective, environmental regimes are driven not only by state powers but also by epistemic networks under certain conditions, such as a knowledge insulated and safeguarded from politics, which means an independent scientific panel; clear procedures connecting science and policy and multiple stakeholders pushing the same direction and supporting the knowledge.

Also, according to Haas' theory, the states recognize the need of dealing with the problem, and they rely on scientists for help only after a widely publicized shock or crisis, as was seen for EBFT for the ICCAT and SBT for the CCSBT. Here, it is all about the decision-makers' recognition of the limits of their abilities to deal with new issues and the need to defer or delegate to authoritative actors with a reputation for expertise. For the CCAMLR, for instance, since science was fully embedded in the institutional design of the agreement, a crisis was not necessary to motivate action.

It is not possible to assure that a policy outcome, influenced by the epistemic community, had a direct effect on the biomass, as it is not possible to control other variables, such as natural variation and IUU control. However, their influence led to policies, which credibly improved biomass. It would be incredibly valuable to study a larger number of agreements which show that when science was listened to, the direct effect promoted a more effective agreement. A quantitative study in this direction would be beneficial.

This book is a contribution to knowledge as it indicates that to improve the complex governance of the oceans, it is important to keep in mind the institutional design for science. When you have agreements, such as the CCAMLR, where sound science and usable knowledge can flow from a strong, trustworthy, and capable epistemic community to the decisions-makers, there exists better management oversight. When decisions are accepted based on the best knowledge available, there is a higher chance of having a more effective agreement, since its alliance is tied to the existence and willingness of key states to exercise leadership, the existence of strong international institutions, the presence of a strong transnational scientific involvement (epistemic communities), and the involvement of NGOs.

Additional research frontiers may include a focus on NGOs and their strategies on international organizations, as well as asking what are the lessons learned about NGO's campaigns for accessing delegates. A process tracing the NGO's actual campaigns to influence delegations would be interesting, or even a description about which NGOs worked with epistemic communities to determine if some NGOs are more science-friendly than others on fisheries management, or if some NGOs act more as advocacy groups following another agenda rather than that offered by scientific knowledge.

As a final note, a focus on epistemic communities and the science and policy interplay to discuss international cooperation do not exclude the central importance of interstate relations in world affairs. However, not to take into account the non-state actors and the role of knowledge in today's world leaves one with only a partial picture of the international system, and thus might represent an incomplete understanding of world politics itself.

Annex I: CCAMLR Epistemic Community

L. R. Gonçalves, *Regional Fisheries Management Organizations*, https://doi.org/10.1007/978-3-030-70362-2

Pais	NAME	Institution	2014	2013	2012	2011	2010	2009	2008	2007	2006	2005	2004	SUM
Convener	Dr Mark Belchier	British Antarctic Survey	0	1	1	1	1	1	0	0	1	1	0	7
Convener	Dr David Agnew	MRAG Ltd	0	0	0	1	1	1	1	1	1	1	1	8
Australia	Dr Andrew Constable	Australian Antarctic Division, Department of the Environment	1	0	0	1	1	1	1	1	1	1	1	9
Australia	Dr Steve Candy	Australian Antarctic Division, Department of the Environment	0	0	1	1	1	1	1	1	1	1	1	9
Australia	Dr Dirk Welsford	Australian Antarctic Division, Department of the Environment	1	1	1	1	1	1	1	1	0	0	0	9
Australia	Dr Philippe Ziegler	Australian Antarctic Division, Department of the Environment	1	1	1	1	1	0	0	0	0	0	0	5
France	Mr Nicolas Gasco	Muséum national d'Histoire naturelle	1	1	1	1	1	1	1	1	1	1	0	10
Germany	Dr Karl-Hermann Kock	Institute of Sea Fisheries—Johann Heinrich von Thünen	1	1	1	0	0	1	1	1	1	1	1	10
Japan	Dr Taro Ichii	National Research Institute of Far Seas Fisheries	1	1	1	0	0	1	0	0	0	0	0	5
Japan	Mr Naohisa Miyagawa	Taiyo A and F Co. Ltd.	0	1	1	1	1	1	1	1	1	0	0	6
Japan	Dr Kenji Taki	National Research Institute of Far Seas Fisheries	1	1	1	1	1	1	0	0	0	0	0	6
New Zealand	Mr Alistair Dunn	National Institute of Water and Atmospheric Research	1	0	0	1	1	1	0	1	1	1	1	8
New Zealand	Mr Jack Fenaughty	Silvifish Resources Ltd.	1	1	1	1	1	1	1	1	1	1	0	10
New Zealand	Dr Stuart Hanchet	National Institute of Water and Atmospheric Research	1	1	1	1	1	1	1	1	1	1	1	11
New Zealand	Dr Steve Parker	National Institute of Water and Atmospheric Research	1	1	1	1	1	1	0	0	0	0	0	7
Russian fed	Dr Konstantin Shust	FSUE-VNIRO	0	0	0	1	1	1	1	1	1	1	1	7
South Africa	Mr Chris Heinecken	Capricorn Fisheries Monitoring (Capfish)	1	0	0	1	1	1	0	1	0	1	0	5

Pais	NAME	Institution	2014	2013	2012	2011	2010	2009	2008	2007	2006	2005	2004	SUM
South Africa	Dr Rob Leslie	Department of Agriculture, Forestry and Fisheries	0	1	1	1	1	1	1	1	0	0	1	8
Spain	Mr Roberto Sarralde Vizuete	Instituto Español de Oceanografía	1	1	1	1	1	0	0	0	0	0	0	5
Ukraine	Dr Leonid Pshenichnov	Methodological and Technological Center of Fishery and Aquaculture	1	1	1	1	1	1	1	1	1	1	1	11
United Kingdom	Dr Martin Collins	Foreign and Commonwealth Office?British Antarctic Survey (2007)	1	1	1	1	1	0	1	1	0	1	1	9
United Kingdom	Dr Rebecca Mitchell	MRAG Ltd	0	0	0	1	1	1	1	1	0	0	0	5
USA	Dr Christopher Jones	National Oceanographic and Atmospheric Administration (NOAA)	1	0	1	1	1	1	1	1	1	1	1	10
USA	Dr Rennie Holt	Southwest Fisheries Science Center	0	0	0	0	0	1	1	1	1	1	1	6
USA	Ms Kim Rivera	National Marine Fisheries Service	0	0	0	0	0	1	1	1	1	1	1	6

Annex II: ICCAT CPCs Year of Ratification

Countries	Year of ratification
United states	1967
Japan	1967
South Africa	1967
Ghana	1968
Canada	1968
France	1968
Brazil	1969
Maroc	1969
Rep of Korea	1970
Cote Divore	1972
Angola	1976
Russia	1977
Gabon	1977
Cap-vert	1979
Uruguay	1983
Sao tome e Principe	1983
Venezuela	1983
Guine Ecuatorial	1987
Guinee rep	1991
United kingdom	1995
Libya	1995
China	1996
European union	1997
Tunisie	1997
Panama	1998
Trinidad and Tobago	1999
Namibia	1999

© The Author(s), under exclusive license to Springer Nature Switzerland AG 2021
L. R. Gonçalves, *Regional Fisheries Management Organizations*, https://doi.org/10.1007/978-3-030-70362-2

Countries	Year of ratification
Barbados	2000
Honduras	2001
Algerie	2001
Mexico	2002
Vanuatu	2002
Iceland	2002
Turkey	2003
Philippines	2004
Norway	2004
Nicaragua	2004
Senegal	2004
Belize	2005
Syria	2005
St Vincent and the Grenadines	2006
Nigeria	2007
Egypt	2007
Albania	2008
Sierra Leone	2008
Mauritania	2008
Curacao	2014
Liberia	2014
El Salvador	2014

Annex III: Epistemic Community at ICCAT

L. R. Gonçalves, *Regional Fisheries Management Organizations*, https://doi.org/10.1007/978-3-030-70362-2

			2004	2005	2006	2007	2008	2009	2010	2011	2012	2013	
		Venue	New Orleans—USA	Seville—Spain	Dubrovnik—Croatia	Antalya—Turkey	Marrakech—Morocco	Recife—Brazil	Paris—France	Istanbul—Turkey	Agadir—Morocco	Cape Town—South Africa	
		COM Chairman	Masanori Miyahara	Masanori Miyahara	W. T. Hogarth	W. T. Hogarth	Fabio Hazin	Fabio Hazin	Fabio Hazin	Fabio Hazin	Masanori Miyahara	Masanori Miyahara	
		SCRS Chairman	J. Gil Pereira	J. Gil Pereira	G. Scott	G. Scott	G. Scott	G. Scott	G. Scott	J. Santiago	J. Santiago	J. Santiago	SUM
1	Brazil	Hazin, Fabio H. V.	1	1	1	1	1	1	1	1	1	1	10
2	Brazil	Travassos, Paulo	1	1	1	1	1	1	1	1	0	1	9
3	Canada	Neilson, John D.	1	1	1	1	1	1	1	1	1	0	9
4	Cape Verde	Marques da Silva Monteiro, Vanda	0	1	1	1	1	1	1	1	1	1	9
5	China, (People's Rep.)	Song, Liming	0	1	1	1	1	1	1	1	1	1	9
6	Croatia	Franicevic, Vlasta	1	1	1	1	1	1	0	1	1	0	8
7	European Community	Ariz Telleria, Javier	1	1	1	1	1	1	0	1	1	1	9
8	European Community	Arrizabalaga, Haritz	1	1	1	1	1	1	1	1	1	1	10
9	European Community	Cort, Jose Luis	1	1	1	1	1	1	1	1	1	1	10
10	European Community	Delgado de Molina Acevedo, Alicia	1	1	1	1	1	1	1	1	1	1	10
11	European Community	Fonteneau, Alain	0	1	1	1	1	1	1	1	1	1	9

			2004	2005	2006	2007	2008	2009	2010	2011	2012	2013	SUM
	Venue		New Orleans—USA	Seville—Spain	Dubrovnik—Croatia	Antalya—Turkey	Marrakech—Morocco	Recife—Brazil	Paris—France	Istanbul—Turkey	Agadir—Morocco	Cape Town—South Africa	
	COM Chairman		Masanori Miyahara	Masanori Miyahara	W. T. Hogarth	W. T. Hogarth	Fabio Hazin	Fabio Hazin	Fabio Hazin	Fabio Hazin	Masanori Miyahara	Masanori Miyahara	
	SCRS Chairman		J. Gil Pereira	J. Gil Pereira	G. Scott	G. Scott	G. Scott	G. Scott	G. Scott	J. Santiago	J. Santiago	J. Santiago	
12	Fromentin, Jean Marc	European Community	1	1	1	1	1	1	1	1	1	1	10
13	Gaertner, Daniel	European Community	1	1	1	1	1	1	1	1	1	0	9
14	Goujon, Michel	European Community	1	1	1	1	1	1	1	1	1	1	10
15	Macías, Ángel David	European Community	1	1	1	1	1	1	1	1	1	1	10
16	Monteagudo, Juan Pedro	European Community	1	1	1	1	1	0	0	1	1	1	8
17	Ortiz de Urbina, Jose Maria	European Community	1	1	1	1	1	1	1	1	1	1	10
18	Ortiz de Zárate Vidal, Victoria	European Community	1	1	1	1	1	1	1	1	1	1	10
19	Pereira, Joao Gil	European Community	1	1	1	1	1	1	1	1	1	0	9
20	Tserpes, George	European Community	1	1	1	1	1	1	1	1	1	1	10
21	Bannerman, Paul	Ghana	0	1	1	1	1	1	1	1	1	1	9
22	Miyake, Makoto P.	Japan	1	1	1	1	1	1	1	1	1	0	9

			2004	2005	2006	2007	2008	2009	2010	2011	2012	2013	
		Venue	New Orleans—USA	Seville—Spain	Dubrovnik—Croatia	Antalya—Turkey	Marrakech—Morocco	Recife—Brazil	Paris—France	Istanbul—Turkey	Agadir—Morocco	Cape Town—South Africa	
		COM Chairman	Masanori Miyahara	Masanori Miyahara	W. T. Hogarth	W. T. Hogarth	Fabio Hazin	Fabio Hazin	Fabio Hazin	Fabio Hazin	Masanori Miyahara	Masanori Miyahara	
		SCRS Chairman	J. Gil Pereira	J. Gil Pereira	G. Scott	G. Scott	G. Scott	G. Scott	G. Scott	J. Santiago	J. Santiago	J. Santiago	SUM
23	Mexico	Ramirez López, Karina	0	1	1	1	1	1	1	1	1	1	9
24	Norway	Nottestad, Leif	0	0	1	1	1	1	1	1	1	1	8
25	Turkey	Karakulak, Saadet	1	1	1	1	1	1	0	0	1	1	8
26	United States	Brown, Craig A.	1	1	1	1	1	1	1	1	1	1	10
27	UnitedOd States	Cass-Calay, Shannon	1	1	1	1	1	1	1	1	1	1	10
28	United States	Cortés, Enric	1	1	1	1	1	1	1	1	0	1	8
29	United States	Díaz, Guillermo	1	1	1	1	1	1	1	0	1	1	9
30	United States	Die, David	1	1	1	1	1	1	1	1	1	1	10
31	United States	Porch, Clarence E.	1	1	1	1	1	1	1	1	1	1	10
32	United States	Prince, Eric D.	1	1	1	1	1	1	1	1	1	1	10
33	United States	Scott, Gerald P.	1	1	1	1	1	1	1	1	1	1	10
34	Uruguay	Domingo, Andrés	0	1	1	1	1	1	1	1	1	1	9

Annex IV: NGOs Attendees at ICCAT

L. R. Gonçalves, *Regional Fisheries Management Organizations*, https://doi.org/10.1007/978-3-030-70362-2

Organizations	2004 PLE	2004 SCRS	2005 PLE	2005 SCRS	2006 PLE	2006 SCRS	2007 PLE	2007 SCRS	2008 PLE	2008 SCRS	2009 PLE	2009 SCRS	2010 PLE	2010 SCRS	2011 PLE	2011 SCRS	2012 PLE	2012 SCRS	2013 PLE	2013 SCRS
Bird Life International	0	0	0	1	0	1	0	0	0	1	1	1	0	0	0	0	0	1	1	0
Oceana	0	0	0	0	0	1	1	0	1	1	1	1	1	0	1	0	0	0	1	1
The ocean conservancy	0	0	0	0	0	1	0	0	1	1	0	0	0	0	0	0	0	0	0	0
WWF	1	0	1	0	1	1	1	1	1	1	1	1	1	1	0	1	1	0	1	1
Fundatun	0	0	0	0	0	1	0	0	0	1	0	0	1	0	0	1	0	0	0	0
Medisamak	0	0	1	1	1	0	1	0	1	1	1	0	1	0	1	0	1	0	1	0
Federation of Maltese Aquaculture Producers (FMAP)	0	0	0	0	0	0	0	0	0	0	1	1	1	1	1	1	1	1	1	1
Greenpeace	0	0	0	0	1	0	1	0	1	0	1	1	1	1	1	1	0	0	1	0
ISSF	0	0	0	0	0	0	0	0	0	0	1	1	1	1	0	0	1	1	1	1
The Pew Environment Group	0	0	0	0	0	0	0	0	1	0	1	1	1	1	1	1	1	1	1	1
Federation of European Aquaculture Producers	0	0	0	0	1	0	1	0	1	0	1	0	1	1	1	1	1	1	0	0
Institute for Public Knowledge	0	0	0	0	0	0	0	0	0	0	0	0	1	0	1	1	1	0	0	0
Conseil Consultatif Regional de la Mediterranee—CCR MED	0	0	0	0	0	0	0	0	0	0	0	0	0	0	1	1	0	1	0	0
Confederation International de la Peche Sportive—CIPS	1	0	1	0	1	0	1	0	1	0	1	0	1	0	1	0	1	1	0	1
IWMC World Conservation Trust	0	0	0	0	0	0	0	0	0	0	0	0	0	0	0	0	0	1	1	0

Organizations	2004 PLE	2004 SCRS	2005 PLE	2005 SCRS	2006 PLE	2006 SCRS	2007 PLE	2007 SCRS	2008 PLE	2008 SCRS	2009 PLE	2009 SCRS	2010 PLE	2010 SCRS	2011 PLE	2011 SCRS	2012 PLE	2012 SCRS	2013 PLE	2013 SCRS
Federcoopesca	0	0	0	0	0	0	0	0	0	0	0	0	0	0	0	0	1	0	0	1
Marine Stewardship Council	0	0	0	0	0	0	0	0	0	0	0	0	0	0	1	0	1	0	0	1
The Ocean Foundation	0	0	0	0	0	0	0	0	0	0	0	0	0	0	0	0	0	0	1	1
Association Euro-Mediterraneenne Des Pecheurs Professionnels De Thon—AEPPT	0	0	0	0	0	0	0	0	0	0	0	0	0	0	1	0	1	0	1	0
Asociación de Pesca, Comercio y Consumo Responsable del Atún Rojo—APCCR	0	0	0	0	0	0	0	0	0	0	0	0	1	0	1	0	1	0	1	0
Blue Water Fishermen's Association	0	0	0	0	0	0	0	0	0	0	0	0	0	0	1	0	1	0	1	0
Defenders of Wildlife	0	0	0	0	0	0	0	0	0	0	0	0	0	0	0	0	0	0	1	0
Ecology Action Centre—EAC	0	0	0	0	0	0	0	0	0	0	0	0	0	0	0	0	1	0	1	0
European Bureau for Conservation and Development—EBCD	0	0	0	0	0	0	0	0	0	0	0	0	0	0	0	0	0	0	1	0
Federpesca	0	0	0	0	0	0	0	0	0	0	0	0	0	0	0	0	1	0	1	0
Organization for Promotion of Responsible Tuna Fisheries—OPRT	0	0	1	0	1	0	1	0	1	0	1	0	1	0	1	0	1	0	1	0
The Varda Foundation	0	0	0	0	0	0	0	0	0	0	0	0	0	0	0	0	0	0	1	0

Organizations	2004		2005		2006		2007		2008		2009		2010		2011		2012		2013	
	PLE	SCRS	PLE	SCRS	PLE	SCRS	PLE	SCRS	PLE	SCRS	PLE	SCRS	PLE	SCRS	PLE	SCRS	PLE	SCRS	PLE	SCRS
Aquatic Release Conservation—ARC	1	0	0	0	0	0	0	0	0	0	0	0	0	0	0	0	0	0	0	0
National Coalition Marine Conservation	1	0	0	0	1	0	0	0	0	0	0	0	0	0	0	0	0	0	0	0
Oceanic Conservation Organization	1	0	0	0	0	0	0	0	0	0	0	0	0	0	0	0	0	0	0	0
Recreational Fishing Alliance—RFA	1	0	0	0	0	0	0	0	0	0	0	0	0	0	0	0	0	0	0	0
Wrigley Institute of ENV Studies—WIES	1	0	1	0	1	0	0	0	0	0	0	0	0	0	0	0	0	0	0	0
International Game Fish Association—IGFA	0	0	0	0	1	0	1	0	0	0	1	0	1	0	1	0	1	0	0	0
Sustainable Fisheries Partnership—SFP	0	0	0	0	0	0	0	0	0	0	0	0	0	0	0	0	0	0	0	0
Association Euroméditerranéenne des Pêcheurs Professionnels de thon—AEPPT	0	0			0	0	0	0	0	0	1	0	1	0	0	0				
Ecology Action Centre (EAC)	0	0	0	0	0	0	0	0	0	0	0	0	0	0	1	0	0	0	0	0
European Bureau for Conservation and Development (EBCD)	0	0	0	0	0	0	0	0	0	0	0	0	1	0	0	0	1	0	0	0
European Elasmobranch Association (EEA)	0	–	0	0	0	0	0	0	0	0	0	0	1	0	0	0	0	0	0	0

Organizations	2004		2005		2006		2007		2008		2009		2010		2011		2012		2013	
	PLE	SCRS	PLE	SCRS	PLE	SCRS	PLE	SCRS	PLE	SCRS	PLE	SCRS	PLE	SCRS	PLE	SCRS	PLE	SCRS	PLE	SCRS
Humane Society International	0	0	0	0	0	0	0	0	0	0	0	0	1	0	0	0	1	0	0	0
IndyACT—The League of Independent Activists	0	0	0	0	0	0	0	0	0	0	0	0	1	0	0	0	0	0	0	0
The Billfish Foundation	0	0	0	0	0	0	0	0	0	0	0	0	1	0	0	0	1	0	0	0
Federation le la Peche Maritime et de la Aquaculture—FPMA	0	0	0	0	0	0	0	0	0	0	0	0	0	0	1	0	1	0	0	0
Natural Resources Defense Council—NRDC	0	0	0	0	0	0	0	0	0	0	0	0	0	0	1	0	0	0	0	0
US Japan Research Inst	0	0	0	0	0	0	0	0	0	0	0	0	0	0	1	0	1	0	0	0
Tuna Producer Association	0	0	0	0	0	0	0	0	0	0	0	0	0	0	0	0	1	0	0	0
SUM	7	0	5	1	9	5	8	1	9	6	13	7	22	6	23	7	23	9	19	9

Annex V

Guide for interviews—not to follow, only for orientation

1. What is your name? And for which Institution do you work?

2. What are the groups you represent: NGOs, University, private sector, and government?

3. How would you describe the events in the specific case (e.g., ICCAT, CCAMLR, CCSBT cases)?

4. In the related case, what are your views of?

5. In your opinion, how science played a role?

6. Who were the stakeholders involved in the case?

7. For decision-makers:

   7.1 At the time of decision-making what is taken into account? How are the political, economic, and scientific views articulated?

   7.2 Are the results of scientific research listened? How do these results come to you?

8. For scientists and Ngos representatives

   8.1 Do you actively participate in an RFMO meeting? Observer? What is your role?

   8.2 Who funds your participation?

   8.3 What is your role/purpose in attending meetings? Is there a work with the national government before?

   8.4 Do you feel that your knowledge is respected by decision-makers? Cite examples. If not, why?

9. Finalization: "Would you like to say something else? Would you like to leave a message?"

© The Author(s), under exclusive license to Springer Nature
Switzerland AG 2021
L. R. Gonçalves, *Regional Fisheries Management Organizations*,
https://doi.org/10.1007/978-3-030-70362-2

References

Adler E (1997) Seizing the middle ground: constructivism in world politics. Eur J Int Rel 3:319–363

Adler E (2002) Constructivism and international relations. In: Handbook of international relations. Sage. http://www.sage-ereference.com/hdbk_intlrelations/Article_n5.html. Accessed 28 Feb 2011

Adler E, Haas PM (1992) Conclusion: epistemic communities, world order, and the creation of a reflective research program. Int Organ 46(1):367–390

Andresen S (2000) Science and Politics in International Environmental Regimes: between Integrity and Involvement. Issues in Environmental Politics. Manchester University Press, New York

Aranda M, De Bruyn P, Murua H (2010) A report review of the tuna RFMOs: CCSBT, IATTC, IOTC, ICCAT and WCPFC. EU FP7 Project (212188), 171

Ardron J, Clark N, Seto K, Brooks C (2013) Tracking twenty-four years of discussions about transparency in international marine governance: where do we stand. Stan Envtl LJ 33:167

ASOC (1988) A statement from ASOC. Eco, XLXI (3), Antarctic and Southern Ocean Coalition (ASOC), Tasmanian Conservation Trust, Hobart, Australia

ASOC (1998) Patagonian toothfish moratorium. Antarctic and Southern Ocean Coalition (ASOC), Washington, USA. http://www.asoc.org/campaign/PattoothfishMoratorium.htm. Accessed 10 Aug 2015

ASOC (2001) The Antarctica project. Antarctic and Southern Ocean Coalition (ASOC), Washington, USA. http://www.asoc.org/. Accessed 7 Dec 2001

ASOC (2002) Pirate fishing: out of control. Eco, ATCM XXV, No 3, Antarctic and Southern Ocean Coalition (ASOC), Tasmanian Conservation Trust, Hobart, Australia

Austral Fisheries (2002) The alphabet boats: a case study of toothfish poaching in the Southern Ocean. Austral Fisheries Pty Ltd, Mt Hawthorn, Australia

Bäckstrand K (2003) Civic science for sustainability: reframing the role of experts, policy-makers and citizens in environmental governance. Global Env Polit 3(4):24–41

Barkin JS, Desombre ER (2013) Saving global fisheries: reducing fishing capacity to promote sustainability. MIT Press

Barrett S (1994) Self enforcing international environmental agreements. Oxf Econ Pap 46:878–894

Beach D, Pedersen RB (2013) Process-tracing methods: foundations and guidelines. University of Michigan Press

Bender P (2007) A state of necessity: IUU fishing in the CCAMLR zone. OCLJ 13:233

Berkman PA, Lang MA, Walton DW, Young OR (2011) Science diplomacy: antarctica, science and the governance of international spaces. Smithsonian Institution Scholarly Press

Betsill MM, Corell E (2001) NGO influence in international environmental negotiations: a framework for analysis. Global Environ Polit 1(4):65–85

L. R. Gonçalves, *Regional Fisheries Management Organizations*, https://doi.org/10.1007/978-3-030-70362-2

Biermann F, Kim R (2020) Architectures of earth system governance: setting the stage. In: Biermann F, Kim R (eds) Architectures of earth system governance: institutional complexity and structural transformation. Cambridge: Cambridge University Press, pp 1–34. https://doi.org/10.1017/9781108784641.001

Biermann F, Pattberg P (eds) (2012) Global environmental governance reconsidered. The MIT Press

Bodin Ö, Österblom H (2013) International fisheries regime effectiveness—activities and resources of key actors in the Southern Ocean. Glob Environ Chang 23(5):948–956

Botcheva L (2001) Expertise and international governance. Glob Gov 7(2):197–224

Boyd D (2002) After the protocol: problems and prospects for Antarctica. In: Jabour-Green J, Haward M (eds) The Antarctic: past, present and future, Antarctic CRC Research Report No. 28. University of Tasmania, Hobart, Australia, pp 105–111

Breitmeier H, Young O, Zurn M (2006) Analyzing international environmental regimes: from case study to database. MIT Press, Cambridge, p 2006

Breitmeier H, Underdal A, Young O (2011) The effectiveness of international environmental regimes: comparing and contrasting findings from quantitative research. Int Stud Rev 13:579–605

Brooks CM (2013) Competing values on the Antarctic high seas: CCAMLR and the challenge of marine-protected areas. Polar J 3(2):277–300

Brooks CM, Weller AJB, Gjerde BK, Sumaila CUR, Ardron DJ, Ban ENC et al (2014) Challenging the right to fish in a fast-changing ocean. Stan Envtl LJ 33:289–457

Butterworth DS, Punt AE, Smith ADM (1996) On plausible hypotheses and their weighting, with implications for selection between variants of the Revised Management Procedure. Rep Int Whaling Commiss 46:637–640

Caron DD (2011) Negotiating our future with the oceans. In: Tubiana L, Jacquet P, Pachauri RK (eds) A planet for life 2011—oceans: the new frontier. pp 25–34. Available at: http://works.bepress.com/david_caron/. Accessed Nov 2012

CCAMLR (1980) CCAMLR convention text. Available at: https://www.ccamlr.org/en/organisation/camlr-convention-text. Accessed Apr 2021

CCAMLR (1988) Seventh meeting of the commission. Hobart, Australia. http://www.ccamlr.org/en/system/files/e-cc-vii.pdf. Accessed 11 Aug 2015

CCAMLR (1996) Report of the working group on fish stock assessment. Hobart, Australia, October 1996. Available at: https://www.ccamlr.org/en/system/files/e-sc-xv-a5.pdf. Accessed May 2021

CCAMLR (1997) Sixteenth meeting of the commission. Hobart, Australia. Available at http://www.ccamlr.org/en/ccamlr-xvi. Accessed 10 Aug 2015

CCAMLR (2008) Performance review panel. Available at https://www.ccamlr.org/en/publications/ccamlr-performance-review. Accessed Dec 2014

CCAMLR (2014) Working group fisheries stock assessment. Available at http://www.ccamlr.org/en/wg-fsa-14 Accessed 10 Aug 2015

CCAMLR (2017) Map of the CAMLR convention area. Last updated October 2017. www.ccamlr.org/node/86816

CCSBT (1994) First meeting of the commission for the conservation of Southern Bluefin Tuna. Wellington, New Zealand

CCSBT (1995a) CCSBT report of the second annual meeting. Tokyo, Japan, 12–15 Sept

CCSBT (1995b) CCSBT report of the first special meeting. Canberra, Australia, 3–6 Oct

CCSBT (1996) CCSBT report of the third annual meeting. Canberra, Australia, 24–28 Sept

CCSBT (1997a) CCSBT report of the resumed third annual meeting (Revised). Canberra, Australia, 18–22 Feb

CCSBT (1997b) CCSBT report of the fourth annual meeting. First Part. Canberra, Australia, 8–13 Sept

CCSBT (1998) CCSBT report of the fourth annual meeting. Second Part. Canberra, Australia, 19–22 Jan

CCSBT (1999a) CCSBT report of the fifth annual meeting. First Part. Tokyo, Japan, 22–26 Feb

CCSBT (1999b) CCSBT report of the sixth annual meeting. First Part. Canberra, Australia, 29, 30 Nov

CCSBT (2000) CCSBT report of the special meeting. Canberra, Australia, 16–18 Nov

CCSBT (2001) CCSBT report of the second stock assessment meeting. Tokyo, Japan, 19–28 Aug

CCSBT (2002) CCSBT report of the first management procedure workshop. Tokyo, Japan, 3–4 and 6–8 Mar

CCSBT (2004) CCSBT report of the fifth stock assessment group. Seogwipo City, Republic of Korea, 6–11 Sept

CCSBT (2005a) CCSBT report of the tenth meeting of the scientific committee. Narita, Japan, 9 Sept

CCSBT (2005b) CCSBT report of the twelfth annual meeting of the commission. Narita, Japan, 15 Oct

CCSBT (2005c) CCSBT Report of the sixth meeting of the stock assessment group. Taipei, Taiwan, 29 Aug–3 Sept

CCSBT (2006a) CCSBT report of the special meeting of the commission. Canberra, Australia, 18–19 July

CCSBT (2006b) CCSBT report of the seventh meeting of the stock assessment group. Tokyo, Japan, 4–11 Sept

CCSBT (2006c) CCSBT report of the eleventh meeting of the scientific committee, Sept

CCSBT (2006d) CCSBT report of the thirteenth annual meeting of the commission. Miyazaki, Japan, 10–13 Oct

CCSBT (2007a) CCSBT Report of the fourteenth annual meeting of the commission. Canberra, Australia, 16–19 Oct

CCSBT (2007b) CCSBT report of the twelfth meeting of the scientific committee. 10–14 Sept

CCSBT (2008) Performance review panel. Available http://www.ccsbt.org/userfiles/file/docs_english/meetings/meeting_reports/ccsbt_15/report_of_PRWG.pdf. Accessed Dec 2014

CCSBT (2011) Report of the eighteenth annual meeting of the commission. Bali, Indonesia, 10–13 Oct

CCSBT (2014) Report of the twenty first annual meeting of the commission. Auckland, New Zealand, 13–16 Oct

Chasek PS (2001) Earth negotiations: analyzing thirty years of environmental diplomacy. United Nations University Press, New York

Clark WC, Majone G (1985) The Critical appraisal of scientific inquiries with policy implications. Sci Technol Hum Values 10(3):6–19

Constable AJ (2011) Lessons from CCAMLR on the implementation of the ecosystem approach to managing fisheries. Fish Fish 12(2):138–151

Constable AJ, William K, Agnew DJ, Everson I, Miller D (2000) Managing fisheries to conserve the Antarctic marine ecosystem: practical implementation of the Convention on the Conservation of Antarctic Marine Living Resources (CCAMLR). ICES J Marine Sci J Conseil 57(3):778–791

Cross DM (2013) Rethinking epistemic communities twenty years later. Rev Int Stud 39:137–160

Cross M (2015) The limits of epistemic communities: EU security agencies. Polit Govern 3(1):90–100

Cullis-Suzuki S, Pauly D (2010) Failing the high seas: a global evaluation of regional fisheries management organizations. Mar Policy 34(5):1036–1042

Darby A (1994) Non-state actors in the Antarctic Treaty System: making heresy orthodox. Cooperative Research Centre for the Antarctic and Southern Ocean Environment (Antarctic CRC) Research Report No 4, Hobart, Australia

De Bruyn P, Murua H, Aranda M (2013) The Precautionary approach to fisheries management: how this is taken into account by Tuna regional fisheries management organisations (RFMOs). Mar Policy 38:397–406

De La Mare WK (1986) Simulation studies on whale management procedures. Thirty sixth Report of the International Whaling Commission. Document SC/37/O14

Desombre ER (2007) The global environment and world politics. Bloomsbury Publishing USA

Dodds K (2000) Geopolitics, Patagonian toothfish and living resource regulation in the Southern Ocean. Third World Q 21(2):229–246

Dunlop CA (2016) Knowledge, epistemic communities, and agenda setting. In: Handbook of public policy agenda setting. Edward Elgar Publishing

Eddie GC (1977) The harvesting of krill. Southern Ocean Fisheries Survey Programme GLO/SO/77/2. Food and Agricultural Organisation, Rome

El-Sayed SZ (1994) History, Organization and Accomplishments of the biomass Programme. In: El-Sayed SZ (ed) Southern ocean ecology: the bio- mass perspective. Cambridge University Press, Cambridge, pp 1–8

Everson I (1977) Antarctic fisheries. Southern Ocean fisheries survey programme. GLO/SO/77/1. Food and Agricultural Organisation, Rome

FAO (2001) International Plan of Action to prevent, deter and eliminate illegal, unreported and unregulated fishing. FAO, Rome, 24 p

FAO (Food and Agriculture Organization) (2004) Excess capacity and illegal fishing: challenges to sustainable fisheries. FAO, Rome, Italy. FAO news story. Available at www.fao.org/newsroom/en/focus/2004/47127/index.html

FAO (2020) The state of world fisheries and aquaculture 2020. Sustainability in action. Rome. https://doi.org/10.4060/ca9229en

Finkelstein LS (1995) What is global governance? Glob Gov 1:367–372

Finnemore M, Sikkink K (2001) Taking stock: the constructivist research program in international relations and comparative politics. Annu Rev Polit Sci 4(1):391–416

Firestone J, Polacheck T (2003) The effectiveness of the UN convention on the law of the sea in resolving international fisheries disputes: the Southern Bluefin Tuna case. In: Harrison NE, Bryner GC (eds) Science and politics in the international environment. Rowman and Littlefield, Lanham, pp 241–270

Fonteneau A (2008) Bilan scientifique et historique de l'ICCAT - Rio de Janeiro 1966-Dubrovnik 2006

Frank RF (1983) Convention on the conservation of Antarctic marine living resources. Ocean Develop Intern Law 13(3):291–345

Fromentin JM, Ravier C (2005) The East Atlantic and Mediterranean bluefin tuna stock: looking for sustainability in a context of large uncertainties and strong political pressures. Bull Mar Sci 76(2):353–362

Garcia SM, Rice J, Charles A (eds) (2014) Governance of marine fisheries and biodiversity conservation: interaction and co-evolution. Wiley

George AL, Bennett A (2005) Case studies and theory development in the social sciences. MIT Press, Cambridge

Gjerde KM, Currie D, Wowk K, Sack K (2013) Ocean in peril: reforming the management of global ocean living resources in areas beyond national jurisdiction. Mar Pollut Bull 74(2):540–551

Gonçalves A, Costa JAF (2011) Governança global e regimes internacionais. Ed Almedina, São Paulo

Grantham GJ (1977) The Utilization of Krill. Southern Ocean fisheries survey programme GLO/SO/77/3. Food and Agricultural Organisation, Rome

Greenpeace (2000) CCAMLR governments are failing the toothfish and albatross. Greenpeace, Amsterdam, The Netherlands. http://archive.greenpeace.org/-oceans/piratefishing/ccamr.html. Accessed 10 July 2002

Grotius H (1604) Mare liberum [the freedom of the seas] (Ralph Van Deman Magoffin trans, Oxford University Press 1916) (1608)

Guzzini S (2000) A reconstruction of constructivism in international relations. Eur J Int Rel 6(2):147–182

Haas EB (1983) Regime decay: conflict management and international organizations, 1945–1981. Int Organ 37(02):189–256

Haas PM (1989) Do regimes matter? Epistemic communities and mediterranean pollution. Int Organ 43(3):377–403

Haas PM (1990) Saving the mediterranean: the politics of international environmental cooperation, the political economy of international change. Columbia University Press, New York

Haas PM (1992) Introduction: epistemic communities and international-policy coordination-introduction. Int Organ 46(1):1–35. Special Edition

Haas PM (2000) Prospects for effective marine governance in the NW pacific region. Mar Policy 24(4):341–348. and "letter to the editor", Marine Policy, 24(6) 499–500

Haas PM (2001) Epistemic communities and policy knowledge. In: Smelser NJ, Wright J, Baltes PB (eds) International encyclopedia of social and behavioral sciences. Elsevier, New York, pp 11578–11586

Haas PM (2004a) science policy for multilateral environmental governance. In: Kanie N, Haas PM (eds) Emerging forces in environmental governance. United Nations University Press, Tokyo, pp 115–136

Haas PM (2004b) When does power listen to truth? a constructivist approach to the policy process. J Eur Publ Policy 11(4):569–592

Haas PM (2006) Evaluating the effectiveness of marine governance. Prepared for the Nippon foundation task force on the dynamics of regional cooperation on oceans and coasts, 34 p

Haas PM (2007) Epistemic communities. In: Bodansky D, Brunnee J, Hey E (eds) The Oxford handbook of international environmental law. Oxford University Press, New York, pp 791–806

Haas PM (2012) Epistemic communities. In: Krieger J (ed) The oxford companion of comparative politics, vol 1. Oxford University Press, Oxford, pp 351–359. https://doi.org/10.1093/acref/9780199738595.001.0001

Haas PM (2014) Reconstructing epistemic communities. Prepared for delivery at the 2014 annual meeting of the American Political Science Association, 28–31 Aug

Haas PM (2015) Epistemic communities, constructivism, and international environmental politics. Routledge, 398 p

Haas PM, Haas EB (2002) Pragmatic constructivism and the study of international institutions. Millennium- J Intern Stud 31(3):573–601

Haas P, Stevens C (2011) Organized science, usable knowledge and multilateral environmental governance. In: Governing the air: the dynamics of science, policy, and citizen interaction, p 125

Haas PM, Keohane RO, Levy MA (1995) The effectiveness of International Environmental Institutions. In: Haas PM, Keohane RO, Levy MA (eds) Institutions for the Earth: sources of effective international environmental protection. MIT Press, Cambridge, pp 3–24

Hanusch F, Biermann F (2020) Deep-time organizations: learning institutional longevity from history. Anthrop Rev 7(1):19–41

Helm C, Sprinz D (2000) Measuring the effectiveness of international environmental regimes. J Confl Resolut 44(5):630–652

Hilborn R, Branch TA, Ernst B, Magnusson A, Minte-Vera CV, Scheuerell MD, Valero JL (2003) State of the world's fisheries. Annu Rev Environ Resour 28(1):359

Hovi J, Sprinz DF, Underdal A (2003) The Oslo-Potsdam solution to measuring regime effectiveness: critique, response, and the road ahead. Global Environ Polit 3(3):74–96

Hurry GD, Hayashi M, Maguire JJ (2008) Report of the independent review. International Commission for the Conservation of Atlantic Tunas (ICCAT). ICCAT, Madrid, 105 p. Available at http://www.iccat.int/Documents/Other/PERFORM_%20REV_TRI_LINGUAL.pdf. Accessed on Oct 2015

IATTC (1950) Convention for the establishment of an inter-american tropical tuna commission. Washington, 1949. Available at: https://www.iattc.org/PDFFiles/IATTC-Instruments/_English/IATTC_IATTC%20Convention%201949.pdf. Accessed May 2021

ICCAT (2004a) Commission report for biennial period, 2004–05. Part I (2004)—vol 1 English version

ICCAT (2004b) Standing Committee on Research and Statistics (SCRS) report for biennial period, 2003–04 PART II (2004)—vol 2 English version

ICCAT (2005a) Commission report for biennial period, 2005–06. Part I (2005)—vol 1 English version

ICCAT (2005b) Standing Committee on Research and Statistics (SCRS) report for biennial period, 2004–05 Part II (2005)—vol 2 English version

ICCAT (2006a) Commission report for biennial period, 2006–07. Part I (2006)—vol 1 English version

ICCAT (2006b) Standing Committee on Research and Statistics (SCRS) report for biennial period, 2005–06 Part II (2006)—vol 2 English version

ICCAT (2007a) Commission report for biennial period, 2007–08. Part I (2007)—vol 1 English version
ICCAT (2007b) Standing Committee on Research and Statistics (SCRS) report for biennial period, 2006–07 Part II (2007)—vol 2 English version
ICCAT (2008a) Commission report for biennial period, 2008–09. Part I (2008)—vol 1 English version
ICCAT (2008b) Standing Committee on Research and Statistics (SCRS) report for biennial period, 2007–08 Part II (2008)—vol 2 English version
ICCAT (2008c) Performance review panel. Available at http://www.iccat.int/Documents/Meetings/Docs/Comm/PLE-106-ENG.pdf. Accessed Dec 2014
ICCAT (2009a) Commission report for biennial period, 2009–10. Part I (2009) —vol 1 English version
ICCAT (2009b) Standing Committee on Research and Statistics (SCRS) report for biennial period, 2008–09 Part II (2009)—vol 2 English version
ICCAT (2010) Commission report for biennial period, 2010–11. Part I (2010)—vol 1 English version
ICCAT (2011a) Commission report for biennial period, 2011–12. Part I (2011)—vol 1 English version
ICCAT (2011b) Standing Committee on Research and Statistics (SCRS) report for biennial period, 2010–11 Part II (2011)—vol 2 English version
ICCAT (2012a) Commission report for biennial period, 2012–13. Part I (2012)—vol 1 English version
ICCAT (2012b) Standing Committee on Research and Statistics (SCRS) report for biennial period, 2011–12 Part II (2012)—vol 2 English version
ICCAT (2013a) Commission report for biennial period, 2013–14. Part I (2013)—vol 1 English version
ICCAT (2013b) Standing Committee on Research and Statistics (SCRS) report for biennial period, 2012–13 Part II (2013)—vol 2 English version
ICCAT (2014a) Commission report for biennial period, 2014–15. Part I (2014)—vol 1 English version
ICCAT (2014b) Standing Committee on Research and Statistics (SCRS) report for biennial period, 2013–14 Part II (2014)—vol 2 English version
ICCAT (2014) Report of the Standing Committee on research and statistics (SCRS) (Madrid, Spain, 29 September to 3 October 2014)
ICCAT (2015) Report of Standing Committee on Research and Statistics (SCRS) (Madrid, Spain, 28 September to 2 October 2015)
Inoue CYA (2003) Regime global de biodiversidade. Comunidades epistêmicas e experiências locais de conservação e desenvolvimento sustentável: o caso Mamirauá. Brasília. Tese (Doutorado em Desenvolvimento Sustentável)–Centro de Desenvolvimento Sustentável, Universidade de Brasília. 348:30–49
IOTC (1993) The agreement for the establishment of the indian ocean Tuna commission. Available at: https://www.iotc.org/about-iotc/basic-texts. Accessed on: April, 2021
ISOFISH (2002) Background longline fishing techniques: Patagonian toothfish profile. International Southern Oceans Longline Fisheries Information Clearing House (ISOFISH), Hobart, Australia, http://www.isofish.org.au/backg/fishing.htm. Accessed 22 Feb 2015
Jasanoff S (2013) A world of experts: science and global environmental constitutionalism. Boston Coll Environ Aff Law Rev 40:439
Jasanoff S, Martello M (eds) (2004) Earthly politics: local and global in environmental governance. MIT Press
Joyner C (2011) Potential challenges to the Antarctic Treaty. In: Berkman PA et al (eds) Science diplomacy: Antarctica, science and the governance of international spaces. Smithsonian Institution Scholarly Press
Keck ME, Sikkink K (1998) Activists beyond borders: advocacy networks in international politics, vol 6. Cornell University Press, Ithaca, NY
Keohane R (1989) International institutions and state power: essays in international relations theory. Westview Press, Boulder

Keohane RO, Nye JS (1987) Power and interdependence revisited. Int Organ 41(04):725–753

Klaer N, Sainsbury K, Polacheck T (1996) A retrospective examination of southern bluefin tuna VPA and projection results, 1982–1995. CCSBT/SC/96/31

Kock KH (2001) The direct influence of fishing and fishery-related activities on non-target species in the Southern Ocean with particular emphasis on longline fishing and its impact on albatrosses and petrels: a review. Rev Fish Biol Fish 11:31–56

Kolody D, Polacheck T, Basson M, Davies C (2008) Salvaged pearls: lessons learned from a floundering attempt to develop a management procedure for Southern Bluefin Tuna. Fish Res 94(3):339–350

Krasner SD (1983) Structural causes and regime consequences: regimes as intervening variables. In: Krasner SD (ed) International regimes. Cornell University Press, Ithaca

Kurota H, Hiramatsu K, Takahashi N, Shono H, Itoh T, Tsuji S (2010) Developing a management procedure robust to uncertainty for southern bluefin tuna: a somewhat frustrating struggle to bridge the gap between ideals and reality. Popul Ecol 52(3):359–372

Kvist S (n.d.) Institutional fragmentation in fisheries management. Available at: http://lup.lub.lu.se/luur/download?func=downloadFile&recordOId=4229163&fileOId=4229168. Accessed 11 Dec 2014

Lack M (2001) Antarctic toothfish: an analysis of management, catch and trade. Traffic Oceania, Sydney, Australia

Lack M, Sant G (2001) Patagonian toothfish: are conservation and trade measures working? Traffic Bull 19(1):1–19

Lidskog R, Sundqvist G (2015) When does science matter? International relations meets science and technology studies. Global Env Polit 15(1):1–20

Lindley J, Techera EJ (2017) Overcoming complexity in illegal, unregulated and unreported fishing to achieve effective regulatory pluralism. Mar Policy 81:71–79

Lodge M (2007) Managing international fisheries: improving fisheries governance by strengthening regional fisheries management organizations. Energ Environ Develop Program 7:1–7

Lodge MW, Anderson D, Løbach T, Munro G, Sainsbury K, Willock A (2007) Recommended best practices for Regional Fisheries Management Organizations: report of an independent panel to develop a model for improved governance by Regional Fisheries Management Organizations. Chatham House, London. 141 p. http://www.oecd.org/dataoecd/2/33/39374297.pdfS

Mackenzie BR, Mosegaard H, Rosenberg AA (2009) Impending collapse of bluefin tuna in the northeast Atlantic and Mediterranean. Conserv Lett 2(1):26–35

Miles EL, Arild U, Steinar A, Jørgen W, Jon Birger S, Carlin EM (2002) Environmental regime effectiveness: confronting theory with evidence. MIT Press, Cambridge, MA

Miller DGM (2011) Sustainable management in the Southern Ocean: CCAMLR science. Science diplomacy: Antarctica, science, and the governance of international spaces, pp 103–121

Milner H (1992) International theories of cooperation among nations: strengths and weaknesses. World Polit 44(03):466–496

Mitchell RB (2003) International environmental agreements: a survey of their features, formation, and effects. Annu Rev Environ Resour 28(1):429–461

Mitchell RB, Clark WC, Cash DW, Dickson NM (eds) (2006) Global environmental assessments: information and influence. MIT Press, Cambridge, MA

Mitchell RB et al (2020) What we know (and could know) about international environmental agreements. Glob Environ Polit 20(1):103–121

Molenaar EJ (2003) Participation, allocation and unregulated fishing: the practice of Regional Fisheries Management Organisations. Int J Marine Coast Law 18:468

Molenaar EJ (2007) Managing biodiversity in areas beyond national jurisdiction. Int J Marine Coast Law 8(1):928

Mooney-Seus ML, Rosenberg AA (2007) Regional fisheries management organizations: progress in adopting the precautionary approach and ecosystem-based management. Chatham House

Mora C, Myers RA, Coll M, Libralato S, Pitcher TJ, Sumaila RU et al (2009) Management effectiveness of the world's marine fisheries. PLoS Biol 7(6):e1000131

Myers RA, Worm B (2003) Rapid worldwide depletion of predatory fish communities. Nature 423:280–283

Najam A, Papa M, Tayab N (2006) Global environmental governance – a reform agenda. International Institute for Sustainable Development, Winnipeg. Available at: http://www.iisd.org/pdf/2006/geg.pdf

OECD (2009) Strengthening regional fisheries management organisations, 129 p

Österblom H, Bodin Ö (2012) Global cooperation among diverse organizations to reduce illegal fishing in the Southern Ocean. Conserv Biol 26(4):638–648

Österblom H, Sumaila UR (2011) toothfish crises, actor diversity and the emergence of compliance mechanisms in the Southern Ocean. Glob Environ Chang 21(3):972–982

Perry M (1998) Australia rallies against Antarctic "fish pirates". Reuters Limited. Available at: http://www.gue.com/news/fishpirates.html Accessed Dec 2014

Peterson MJ (1992) Whalers, cetologists, environmentalists, and the international management of whaling. Int Organ 46(01):147–186

Pielke RA Jr (2004) When scientists politicize science: making sense of controversy over. Skeptical Environment Environmental Sci Policy 7(5):405–417

Polacheck T (2002) Experimental catches and the precautionary approach: the Southern Bluefin Tuna Dispute. Mar Policy 26:283–294

Polacheck T (2012) Politics and independent scientific advice in RFMO processes: a case study of crossing boundaries. Mar Policy 36(1):132–141

Polacheck T, Eveson P (2007) Analyses of tag return data from the CCSBT SRP tagging program – 2007. Commission for the conservation of southern bluefin tuna working paper CCSBT-ESC/0709/19

Polacheck T, Klaer NL, Millar C, Preece AL (1999) An initial evaluation of management strategies for the southern bluefin tuna fishery. ICES J Mar Sci 56:811–826

Riddle KW (2006) Illegal, unreported, and unregulated fishing: is international cooperation contagious? Ocean Develop Intern Law 37(3–4):265–297

Ridgeway L (2014) Global level institutions and processes: frameworks for understanding critical roles and foundations of cooperation and integration. In: Garcia SM, Rice J, Charles A (eds) Governance of marine fisheries and biodiversity conservation: interaction and co-evolution. Wiley

Rosenau JN (1992) Governance, order, and change in world politics. In: Rosenau J, Czempiel E-O (eds) Governance without government: order and change in world politics. Cambridge University Press, Cambridge, pp 1–29

Ruckelshaus M, Klinger T, Knowlton N, Demaster DP (2008) Marine ecosystem-based management in practice: scientific and governance challenges. Bioscience 58:53–63

Ruggie JG (1975) International responses to technology: concepts and trends. Int Organ 29(03):557–583

Ruggie JG (1998) What makes the world hang together? Neo-utilitarianism and the social constructivist challenge. Int Organ 52:855–887

Sainsbury K, Polacheck T (1993) Development of a stock rebuilding strategy for Southern Bluefin Tuna. In: 12th SBT trilateral meeting scientific meeting, SBFWS/93/19, Hobart, Australia, 13–19 Oct

Scholte JA (2004) Civil society and democratically accountable global governance. Gov Oppos 39(2):211–233

Schroeder H, Leslie AK, Tay S (2008) Contributing to the science–policy interface: policy relevance of findings on the institutional dimensions of global environmental change. In: Young O, King LA, Schroeder H (eds) Institutions and environmental change: principal findings, applications, and research frontiers. MIT Press, Cambridge, MA, pp 261–275

Shusterich KM (1984) The Antarctic Treaty System: history, substance, and speculation. Int J 39:800–827

Smith ADM, Sainsbury KJ, Stevens RA (1999) Implementing effective fisheries- management systems—management strategy evaluation and the Australian partnership approach. ICES J Mar Sci 56:967–979

Snidal D (1993) Relative gains and the pattern of international cooperation. In: Baldwin DA (ed) Neorealism and neoliberalism: the contemporary debate. Columbia University Press, New York, pp 170–208

Social Learning Group (2001) Learning to manage global environmental risks: a comparative history of social response to climate change, ozone depletion, and acid rain. MIT Press, Cambridge, MA

Speth JG, Haas PM (2006) Global environmental governance. Island Press, Washington, DC

Sumaila UR, Huang L (2012) Managing bluefin tuna in the Mediterranean Sea. Mar Policy 36(2):502–511

Sumaila UR, Zeller D, Watson R, Alder J, Pauly D (2007) Potential costs and benefits of marine reserves in the high seas. Mar Ecol Prog Ser 345:305–310

Toke D (1999) Epistemic communities and environmental groups. Politics 19(2):97–102

TRAFFIC (2001) Patagonian toothfish: are conservation measures working? Illegal fishing continues to threaten Patagonian toothfish. Trade Records Analysis of Flora and Fauna in International Commerce (TRAFFIC). http://www.traffic.org/toothfish/tooth2.html. Accessed 10 Aug 2015

TRAFFIC, WWF (2002) A CITES priority: TRAFFIC and WWF briefing document. World Patagonian toothfish and Antarctic toothfish at the twentieth meeting of the Conference of Parties to CITES (COP12), Santiago, Chile, 2002, Trade Records Analysis of Flora and Fauna in International Commerce (TRAFFIC) and World Wildlife Fund (WWF), Surrey, UK

Underdal A (1992) The concept of regime effectiveness. Coop Confl 27(3):227–240

Veiga JE (2013) A desgovernança mundial da sustentabilidade. Editora 34. ISBN 978-85-7326-518-9

Viola E, Franchini M, Ribeiro TL (2013) Sistema Internacional de Hegemonia Conservadora: Governança Global e Democracia na Era da Crise Climática, Ed Annablume, ISBN 978–85–391-0506-9

Voigt S (2013) How (Not) to measure institutions. J Inst Econ 9:1–26

Wapner P (2000) The transnational politics of environmental NGOs: governmental, economic, and social activism. In: Chasek PS (ed) The global environment in the twenty-first century: prospects for international cooperation. United Nations University Press, New York, pp 87–108

Watson RT (2005) Turning science into policy: challenges and experiences from the science–policy interface. Philos Trans R Soc B Biol Sci 360(1454):471–477

WCPFC (2004) Convention on the conservation and management of high migratory fish stocks in the western and central pacific ocean. Available at: https://www.wcpfc.int/convention-text. Accessed on: April, 2021

Webster DG (2009) Adaptive governance: the dynamics of Atlantic fisheries management. MIT Press

Webster DG (2011) The irony and the exclusivity of Atlantic Bluefin Tuna management. Mar Policy 35(2):249–251

Willock A, Lack M (2006) Follow the leader: learning from experience and best practice in regional fisheries management organizations. Available at http://wwf.panda.org/about_our_earth/blue_planet/publications/?69480/Follow-the-leader-Learning-from-experience-and-best-practice-in-regional-fisheries-management-organizations. Accessed Dec 2014

Young OR (1989) International cooperation: building regimes for natural resources and the environment. Cornell University Press

Young OR (1999) The effectiveness of international environmental regimes: causal connections and behavioral mechanisms. MIT Press, Cambridge, MA

Young OR (2003) Environmental governance: the role of institutions in causing and confronting environmental problems. Int Environ Agreements Polit Law Econ 3:377–393. https://doi.org/10.1023/b:inea.0000005802.86439.39

Young OR, Levy MA (1999) The effectiveness of international environmental regimes. In: The effectiveness of international environmental regimes: causal connections and behavioral mechanisms. MIT Press, Cambridge, MA

Index

Printed by Printforce, the Netherlands